Television and Teletext

Macmillan New Electronics Series
Series Editor: Paul A. Lynn

Paul A. Lynn, *Radar Systems*
A. F. Murray and H. M. Reekie, *Integrated Circuit Design*
Dennis N. Pim, *Television and Teletext*
Martin S. Smith, *Introduction to Antennas*

Television and Teletext

Dennis N. Pim

BSc (Eng), PhD, C Eng, MIEE
Discipline of Electronics, Faculty of Technology, The Open University

Macmillan New Electronics
Introductions to Advanced Topics

© Dennis N. Pim 1988

All rights reserved. No reproduction, copy or transmission of this publication may be made without written permission.

No paragraph of this publication may be reproduced, copied or transmitted save with written permission or in accordance with the provisions of the Copyright Act 1956 (as amended), or under the terms of any licence permitting limited copying issued by the Copyright Licensing Agency, 33-4 Alfred Place, London WC1E 7DP.

Any person who does any unauthorised act in relation to this publication may be liable to criminal prosecution and civil claims for damages.

First published 1988

Published by
MACMILLAN EDUCATION LTD
Houndmills, Basingstoke, Hampshire RG21 2XS
and London
Companies and representatives
throughout the world

Printed in Hong Kong

British Library Cataloguing in Publication Data
Pim, Dennis N.
   Television and teletext.—(Macmillan
   new electronics).
   1. Television   2. Teletext (Data
   transmission system).
   I. Title
   621.388   TK6630

ISBN 0-333-45098-1
ISBN 0-333-45099-X Pbk

Contents

Series Editor's Foreword   ix
Preface   x

1 Monochrome Television   1
  1.1 Light and colour   1
  1.2 Converting scenes to electrical signals   3
    1.2.1 Visual acuity   3
    1.2.2 Dividing up a scene   4
  1.3 Scanning   4
    1.3.1 Number of lines   7
    1.3.2 Scanning rate   10
    1.3.3 Interlaced scanning   11
  1.4 Gamma correction   14
  1.5 Synchronisation   15
  1.6 The composite video signal   15
    1.6.1 Bandwidth of the composite video signal   19
  1.7 Transmitting the video signal   21
  1.8 System variations   23
  1.9 Summary   24

2 Colour Television   26
  2.1 The three colour signals   26
  2.2 Compatibility   27
  2.3 Producing and displaying the colour signals   28
    2.3.1 Colour television tubes   28
  2.4 Colour difference signals   29
  2.5 The NTSC signal   31
    2.5.1 Monochrome signal   32
    2.5.2 Chrominance signals   32
    2.5.3 Luminance/chrominance interference effects   35
    2.5.4 Sub-carrier burst   37
    2.5.5 Choice of sub-carrier frequency   38
    2.5.6 Summary   40
  2.6 The PAL system   42
    2.6.1 Effects of sub-carrier phase errors in the NTSC system   42

	2.6.2 Colour difference bandwidths	44
	2.6.3 Phase reversal	44
	2.6.4 Simple PAL	45
	2.6.5 Delay line PAL	46
	2.6.6 Swinging burst	49
	2.6.7 Sub-carrier frequency	49
	2.6.8 Summary	50
2.7	The SECAM system	50
	2.7.1 Chrominance signal weighting	52
	2.7.2 Video pre-emphasis	52
	2.7.3 FM modulation	53
	2.7.4 Sub-carrier pre-emphasis	54
	2.7.5 Chrominance identification signal	55
	2.7.6 Interference and compatibility effects	56
	2.7.7 Summary	56
2.8	Summary of the three colour encoding systems	56

3 Teletext — 58

3.1	The basic teletext system	58
3.2	Service organisation	60
	3.2.1 Page addressing	61
3.3	The teletext display	61
	3.3.1 Determination of character size	61
	3.3.2 Display characters	62
	3.3.3 Character coding	63
3.4	The teletext TV line	66
	3.4.1 Data encoding	66
	3.4.2 Decoder synchronisation	67
	3.4.3 Addressing data	68
	3.4.4 Hamming error protection	69
	3.4.5 Magazine and row address group	70
	3.4.6 Teletext display packets 1–23	71
	3.4.7 Header row format	71
	3.4.8 Header control bits	73
3.5	The teletext signal	73
	3.5.1 Bit rate	76
	3.5.2 Bandwidth	76
	3.5.3 Data rate	77
3.6	Full channel teletext	78

4 Videotex — 79

4.1	The basic videotex system	80
4.2	The videotex database	81
	4.2.1 Tree structure	81

		4.2.2 Charging structure	81
	4.3	The videotex display	82
		4.3.1 Videotex control codes	83
	4.4	The videotex signal	83
		4.4.1 Decoding the serial signal	85
		4.4.2 Signal modulation	85
		4.4.3 Data rate	86
5	**Television Receivers**		**87**
	5.1	Reception and demodulation of the video signal	88
		5.1.1 Tuner unit	89
		5.1.2 Frequency synthesis tuning	90
		5.1.3 IF signal processing	91
	5.2	Audio signal processing	92
	5.3	Synchronisation signal processing	93
		5.3.1 Sync separator	94
		5.3.2 Line oscillator and waveform shaping	95
		5.3.3 Field oscillator and waveform shaping	96
		5.3.4 Scan signal amplifiers	96
	5.4	Video signal processing	97
		5.4.1 NTSC colour decoder	98
		5.4.2 PAL colour decoder	99
		5.4.3 SECAM colour decoder	100
	5.5	Teletext decoder	101
		5.5.1 Sync and timing generator	101
		5.5.2 Data signal conditioning	102
		5.5.3 Data clock generation	103
		5.5.4 Serial-to-parallel conversion	103
		5.5.5 Main control logic	103
		5.5.6 Display memory	104
		5.5.7 Display unit	104
	5.6	Videotex decoder	105
		5.6.1 Telephone line barrier	106
		5.6.2 Modem	106
		5.6.3 Controller	106
		5.6.4 Display memory and controller	107
	5.7	Control module	107
		5.7.1 Remote control transmitter	109
		5.7.2 Remote control receiver	109
		5.7.3 Front panel controls	109
		5.7.4 Main control unit	109
		5.7.5 Control bus	110
		5.7.6 Bus receivers	110
	5.8	Power supply	110

Contents

5.9	Periconnector	111
5.10	Summary	111

6 New Specifications — **113**

- 6.1 Stereo digital sound — 113
  - 6.1.1 Choice of method — 113
  - 6.1.2 Analogue-to-digital conversion — 114
  - 6.1.3 Data frames — 115
  - 6.1.4 Modulation method — 117
  - 6.1.5 Summary — 118
- 6.2 The MAC system — 118
  - 6.2.1 The basic MAC system — 119
  - 6.2.2 The MAC variants — 120
  - 6.2.3 Synchronisation — 121
  - 6.2.4 Control data — 121
  - 6.2.5 Time compression — 122
  - 6.2.6 Sound coding — 123
  - 6.2.7 C-MAC data packets — 124
  - 6.2.8 Scrambling — 126
  - 6.2.9 Summary — 127
- 6.3 The World System Teletext specification — 127
  - 6.3.1 The need for enhancements — 127
  - 6.3.2 525-line systems — 128
  - 6.3.3 Levels — 128
  - 6.3.4 Implementation — 131
  - 6.3.5 Progress on implementation — 137
  - 6.3.6 Summary — 138
- 6.4 The CEPT Videotex specification — 139
  - 6.4.1 Presentation Protocol Data Units — 139
  - 6.4.2 Summary — 142
- 6.5 Overview — 142

Further reading — 143

Index — 145

Series Editor's Foreword

The rapid development of electronics and its engineering applications ensures that new topics are always competing for a place in university and polytechnic courses. But it is often difficult for lecturers to find suitable books for recommendation to students, particularly when a topic is covered by a short lecture module, or as an 'option'.

Macmillan New Electronics offers introductions to advanced topics. The level is generally that of second and subsequent years of undergraduate courses in electronic and electrical engineering, computer science and physics. Some of the authors will paint with a broad brush; others will concentrate on a narrower topic, and cover it in greater detail. But in all cases the titles in the Series will provide a sound basis for further reading of the specialist literature, and an up-to-date appreciation of practical applications and likely trends.

The level, scope and approach of the Series should also appeal to practising engineers and scientists encountering an area of electronics for the first time, or needing a rapid and authoritative update.

Paul A. Lynn

Preface

This book is about the fundamental systems which enable pictures and text to be converted to electrical signals, transmitted, received and converted back to pictures and text. It is not a book on how to design television cameras, transmitters, receivers etc., but seeks only to introduce the basic mechanisms by which pictures and text can be electronically represented. One of the main aspects of the discussion is on how the characteristics of the physical quantity being represented (pictures, sound, written word) influence the way in which it is represented by electrical signals.

The book is designed for undergraduates who require a broad knowledge as well as technical students who require a basic understanding of the theory behind television and teletext systems. Practising engineers should also find the book valuable. Although the systems adopted by Britain are discussed in a little more detail than the other systems in use around the world, these other systems are explained in some depth, and comparisons made between them. The book should thus be useful for anyone who requires an introduction to television.

Space does not allow a detailed discussion of all the various systems in use today, but it is hoped that sufficient detail is given so that the reader can grasp the basic ideas and understand why the systems have developed in the way in which they have. For further reading on any of the topics discussed, the reader is referred to the Further Reading section at the end of the book, as well as to the published full specifications of each system.

The level of the book assumes a basic knowledge of analogue and digital techniques and the basic properties of light, sound and text. Specifically a knowledge of the following topics is assumed:

- the conversion of light to/from an electrical signal (photocell/bulb)
- the conversion of sound to/from an electrical signal (microphone/loudspeaker)
- the cathode ray tube
- AM and FM modulation
- frequency and bandwidth
- binary representation of information (ASCII codes)
- synchronous and asynchronous serial data transmission

Chapter 1 looks at what light is, and shows how a picture can be divided up into a series of strips and then converted into an electrical signal. The basic black and white composite video signal is then developed and the various line systems introduced. The final section looks at how the signal is broadcast in the UHF and VHF radio bands.

Chapter 2 explains colour, and how we perceive it. It is then shown how the black and white composite video signal can be modified to include colour information, while still retaining compatibility. The three main colour encoding systems — NTSC, PAL and SECAM — are described and their relative merits compared.

Chapter 3 describes the original teletext system as developed in Britain in the early 1970s, and now adopted and in use in many countries throughout the world.

Chapter 4 looks at the basic videotex system and shows how what is basically a computer time-share system has been modified to provide a consumer-oriented system that employs the teletext display to enable a suitably equipped television to be used as a terminal.

Chapter 5 describes the basic operation of the various sections of television receivers that are equipped to receive and display teletext and videotex as well as colour pictures. This chapter does not explain how to design a television set, but seeks to show the various stages that the received signal undergoes in order to produce pictures, sound and teletext data.

Chapter 6 briefly introduces the features and facilities of some of the new specifications that are currently being finalised in the field of television and teletext. Most of these new systems will be appearing over the next few years, and the chapter shows how advances in electronic technology have enabled such features and facilities to become a practical and economic possibility. The complexity of each of these new specifications means that only an overall discussion of each one is possible. Also, in some cases further detail would be inappropriate since some of the specifications have, at the time of writing not yet been finalised and internationally agreed.

Dennis N. Pim

The Open University
Milton Keynes
Buckinghamshire MK7 6AA

Acknowledgements

The author and publishers wish to thank the following who have kindly given permission for the use of copyright material.

British Airways Plc. for Figure 6.11.
Oracle Teletext Ltd for Figures 3.4 and 6.16.
Unwin Hyman Ltd for Figure 2.3 from *Telecommunications: A Systems Approach* by G. Smol, M. P. R. Hamer and M. T. Hills, Allen and Unwin/ The Open University, Figure 6.10.

Every effort has been made to trace all the copyright holders but if any have been inadvertently overlooked the publishers will be pleased to make the necessary arrangement at the first opportunity.

1 Monochrome Television

This chapter will look at the various components that go to make up the television signal which is used to transmit black and white (*monochrome*) pictures. Chapter 2 will show how this signal can be modified to enable colour pictures to be transmitted.

1.1 Light and colour

Light consists of high-frequency electromagnetic waves. The band of frequencies which we perceive as light forms a small part of a whole spectrum of electromagnetic waves which encompass gamma-rays, X-rays, ultra-violet, infra-red (heat), radar and radio waves. The different frequencies within the visible light band are perceived as different colours. The eye responds to wavelengths of electromagnetic waves in the range of about 380–710 nm (1 nm (nano-metre) = 10^{-9} metres).

Figure 1.1 shows the approximate wavelengths of the colours in the visible spectrum, and figure 1.2 shows the wavelengths of the whole electromagnetic spectrum.

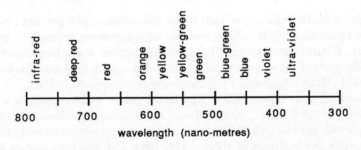

Figure 1.1 The visible light spectrum

The eye's response is not constant with frequency, being about 100 times more sensitive to yellow–green light in the middle of the range as it is to light at the edges of the range. This response is called the *visibility function* or the *photopic response*. The normal photopic response is shown in figure 1.3.

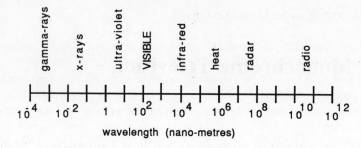

Figure 1.2 The electromagnetic spectrum

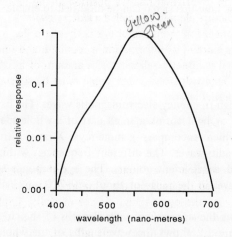

Figure 1.3 Photopic response

Any particular frequency of light within the visible range is perceived by us as having a particular colour. Such colours are known as *monochromatic* or *spectral colours*. It turns out that most light sources produce many frequencies from different parts of the spectrum, but we still perceive each such source as having a particular colour — known as a *polychromatic* or *non-spectral colour*.

Therefore, what is perceived as a particular colour is not necessarily a single frequency of light. Indeed, many different combinations of colours, spectral and non-spectral, can be combined to produce the same perceived colour. In particular, suitable combinations of spectral red, green and blue light sources can be used to create the same visual effect as most of the colours which we see. The spectral colours red, green and blue, therefore, are known as *primary* colours, and it is this property that is used to effect colour television.

As an example, light from the sun which we perceive as white light consists of all frequencies within (and outside) the visible spectrum. However, when we look at a white scene on a colour television, the light is composed only of suitable proportions of red, green and blue light frequencies.

1.2 Converting scenes to electrical signals

In its simplest form, a photocell can be used to convert light intensity into a varying electrical signal, and an ordinary light bulb can be used for the reverse process. But, if we wish to convert a complete scene to/from electrical signals, such a system would not be satisfactory, since a single photocell/bulb combination would only convey a representation of the average light intensity of the whole scene.

What is required is for the scene to be split up into a large number of small areas, and the intensity (and colour) of the light in each of these areas transmitted and then recreated in its original relative position at the receiving end. A signal composed of such picture elements is called a *video signal*.

An important conclusion from the above is that in order to transmit a complete picture, a large amount of information is required, and since the bandwidth of an electrical signal is directly related to the amount of information the signal is carrying, so the bandwidth of a video signal must be substantial — it will be seen just how substantial later.

1.2.1 Visual acuity

To get a fair representation of an image, the size of the picture elements into which it must be divided depends on the amount of detail the eye can perceive, or the *visual acuity* of the eye. A high degree of acuity implies the ability to distinguish objects which are very close together. Hence visual acuity, V, is defined as the reciprocal of the angle, θ (in radians), that an object subtends at the observer's eye:

$$V = \frac{1}{\theta} \tag{1.1}$$

This is illustrated in figure 1.4.

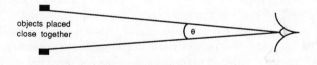

Figure 1.4 Measuring visual acuity

Visual acuity depends on a number of factors including the shape of objects, the intensity of the light and the contrast between the objects and their background. A value of 3400 radians^{-1} has been generally accepted from the results of many tests on viewers under normal television viewing conditions.

1.2.2 Dividing up a scene

How can a scene be converted to electrical signals? One method is to have an array of photocells on to which the scene is focused. Each photocell is then interrogated electrically, in turn, and its output transmitted. At the receiving end there is a similar array of light bulbs and the output from each photocell is connected to the equivalent light bulb. Such a system is illustrated in figure 1.5.

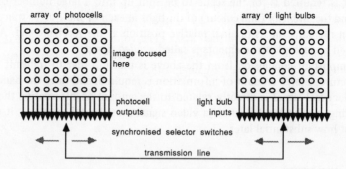

Figure 1.5 A photocell/light bulb array

Because of technical limitations on the number and size of photocell arrays that can be produced, a second, more common way is to scan the picture in a pre-determined pattern and transmit a signal representing the instantaneous brightness of that part of the picture as the scan progresses. If, at the receiving end, a light source is made to scan the display screen in the same pattern and at the same time, and the intensity of the source is varied according to the received electrical signal, then the picture should be reproduced.

These, then, are the elements of a television system which will now be discussed in more detail.

1.3 Scanning

Two methods of dividing up a scene have been mentioned. The first, using an array of photocells, seems very complicated, since to get sufficient resolution, a large number of devices must be used, and each one has to be able to be individually addressed. However, developments in integrated circuit technology have led to arrays of light-detecting devices being able to be integrated on to a single silicon chip complete with the necessary address decoding circuitry. As yet these devices usually have only a reduced resolution, and are thus becoming used in such applications as security cameras etc. where a high resolution is not required.

Monochrome Television

The second method, where the picture is scanned, is called *scanning* and is the basis of how television systems operate.

The vast majority of television cameras still effect the scanning process using an electron beam. In this type of camera the image is focused on to a light-sensitive surface which produces a charge pattern on the surface. The amount of charge is related to the amount of light at each point. An electron beam is made to scan the image from left to right and top to bottom by being deflected simultaneously in the vertical and horizontal directions. The light-sensitive surface and electron beam form an electrical circuit which includes a load resistor. As the beam scans the surface, variations in the charge on the surface cause variations in the current flow round the circuit. The voltage across the load resistor thus varies, and it is this voltage which is the video output signal. Figure 1.6 shows a simplified diagram of a TV camera tube.

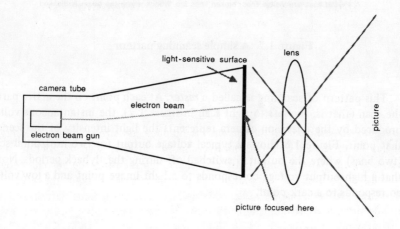

Figure 1.6 Simplified diagram of a TV camera tube

Figure 1.7 shows a simple scanning pattern. The horizontal motion is called the *line scan* and the vertical motion is called the *field scan*. The left to right motions in the field scan are the active picture lines which actually scan the image, and the right to left motions are the much shorter periods when the electron beam is returning to the left of the screen ready for the next active line scan. These periods are known as *line flybacks*. Similarly, the time when the electron beam is returning to the top of the screen at the end of one field scan, ready for the next, is the *field flyback* interval. In figure 1.7, this interval takes two line scan intervals to complete. Notice that the horizontal line scan motion never stops, even during field flyback.

Television and Teletext

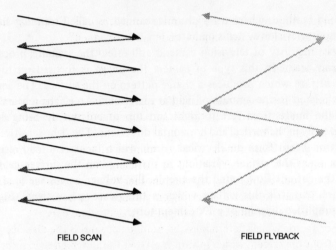

Figure 1.7 A simple scanning pattern

This pattern of scanning is called a *raster*. At each point on the active part of the scan (that is, the left to right scan in figure 1.7), the instantaneous voltage produced by the television camera represents the light intensity of the scene at that point. Figure 1.8 shows a typical voltage output for two horizontal scans (two lines) where the output is switched off during the flyback periods. Notice that a high output voltage corresponds to a light image point and a low voltage corresponds to a dark point.

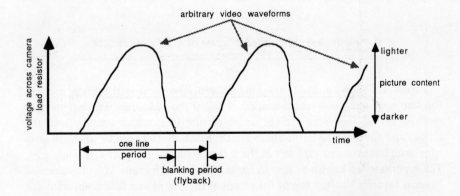

Figure 1.8 Typical video signal from a TV camera for two line periods

Monochrome Television

For a practical system there are two main considerations: how many line scans should there be, and how fast should the image be scanned?

1.3.1 Number of lines

The number of lines is determined by the acuity of the eye. The fewer the number of lines, the less clearly will the picture be defined and the more individual scans will become visible.

How can an estimate for the required number of scans be determined? The initial assumption is that there should be sufficient lines for each individual line to be just not detectable. Or to put it another way, the number of scan lines should be sufficient to resolve a picture containing a set of horizontal bars spaced such that the eye is just unable to distinguish the individual bars.

Another factor which determines the number of lines is the distance that the viewer is from the screen, which in turn is related to the screen size. Experiments have shown that the average viewing distance is about eight times the height of the screen.

Figure 1.9 shows a set of line scans on a television picture of height H where each scan is a distance d apart (that is, there are H/d line scans). If the viewer sits a distance D from the screen, then the angle θ subtended from the eye to each consecutive line scan will be $\approx d/D$ radians (since $\sin \theta = d/D$ and if θ is small $\sin \theta \approx \theta$).

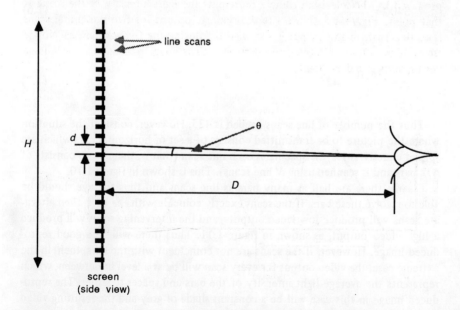

Figure 1.9 Determining the vertical resolution

From above, the visual acuity of the eye, V, is given by

$$V = \frac{1}{\theta} \tag{1.2}$$

hence

$$V = \frac{D}{d} \tag{1.3}$$

Also the number of line scans required, N, is given by

$$N = \frac{H}{d} \tag{1.4}$$

hence

$$V = \frac{DN}{H} \tag{1.5}$$

The generally accepted value of acuity is 3400 radians^{-1}, and also $D \approx 8H$. Hence

$$3400 = \frac{8HN}{H} \tag{1.6}$$

or

$$N = \frac{3400}{8} = 425 \tag{1.7}$$

Thus the number of line scans needed is 425. However, consider the situation where the picture to be transmitted consists of a set of horizontal bars which are spaced exactly twice the spacing as each line scan (that is, the picture consists of $N/2$ bars and is scanned using N line scans). This is shown in figure 1.10.

Clearly there are half as many bars as line scans and thus the eye should be able to resolve these bars. If the scans exactly coincide with the bars, then alternate scans will produce low video outputs, and the intervening scans will produce a high video output, as shown in figure 1.11. Thus there will be a good reproduced image. However, if the scans are not coincident with the bars, then, in the extreme case, the video output for every scan will be at a level in between, which represents the average light intensity of the bars and spaces together. The reproduced image in this case will be a constant shade of grey and the resulting video waveform will be as shown in figure 1.12. Thus, under some circumstances, even the calculated value of 425 lines will not be sufficient.

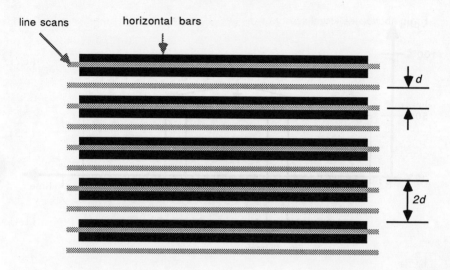

Figure 1.10 Enlarged view of the scanning of a set of narrowly spaced bars

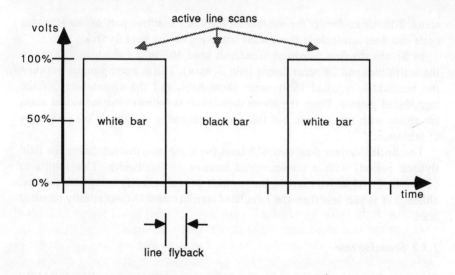

Figure 1.11 Video output from coincident scanning of narrow bars

Statistical experiments have shown that to cater for this type of display, the number of line scans should be increased by a factor of $1/K$ where K is the *Kell factor*. This factor is usually about 0.7, hence to be able to resolve a horizontal bar display of the type described above there needs to be 425/0.7 = 607 line

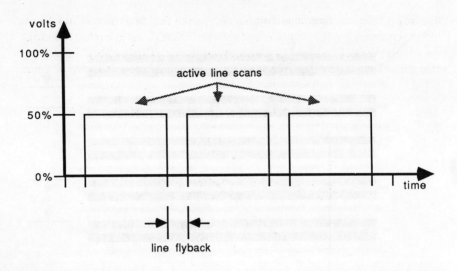

Figure 1.12   Video output from non-coincident scanning of narrow bars

scans. This, of course, is the number of lines in the active part of the scanning cycle and does not include those scans occurring during field flyback.

In Britain the first television broadcasts used 405 lines (of which 377 are in the active area and 28 occur during field flyback). This is a compromise between the bandwidth required to transmit these lines, and the acceptability of the reproduced picture. From the above derivations it follows that individual scans are visible with this system, but this was considered acceptable in the early days of television.

The British system now uses 625 lines (with 40 lines unused during the field flyback period) with a consequential increase in bandwidth. This results in 625 − 40 = 585 active scanning lines, which gives a perfectly acceptable picture, although it is just less than the calculated number of 607 theoretically required lines.

1.3.2 Scanning rate

The second consideration is the rate of scanning — that is, how fast should the electron beam be made to traverse the screen?

In the television receiver, as the electron beam hits the screen, it causes a dot of light to appear at that point. After the beam has left the point, the light decays rapidly (but not instantaneously). The scanning process therefore causes each area of the screen to produce short light pulses with much longer intervening dark periods. If this interval is too long the picture will appear to flicker. This is known as the *flicker effect*, and observations have shown that this is dependent not only on the scanning frequency but also on the spot brightness. This effect is offset somewhat by the persistence of the eye – that is, the eye retains an image for a short period. Experimental results, averaged over a large number of observers, give typical results for the flicker effect *vs* spot brightness as shown in figure 1.13.

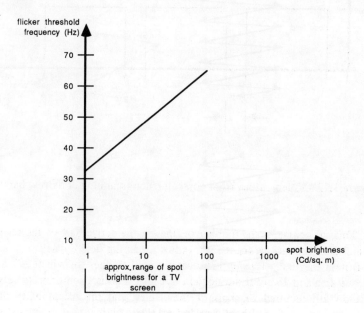

Figure 1.13 Flicker effect *vs* spot brightness

From this figure, it appears that for the normal range of brightnesses on a TV screen, a rate of about 60 scans per second is required if the flicker effect is not to be noticeable. However, because the rate of scanning determines the bandwidth of the video signal, if the bandwidth can be reduced without increasing the apparent flicker, then so much the better. This can be achieved by interlaced scanning.

1.3.3 Interlaced scanning

Consider the situation where instead of the electron beam scanning every line from the top to the bottom of the screen in one go, it scanned every alternate line in the first scan, and 'filled in' the missing lines in the second scan. The overall effect is that the beam scans the picture twice before every line has been displayed but, after each scan, a complete picture has been produced, using only half the total number of lines. This is known as *interlaced scanning*, and is illustrated in figures 1.14 and 1.15.

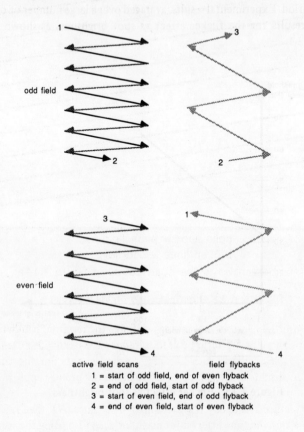

active field scans   field flybacks
1 = start of odd field, end of even flyback
2 = end of odd field, start of odd flyback
3 = start of even field, end of odd flyback
4 = end of even field, start of even flyback

Figure 1.14   Simple interlace scanning

Interlaced scanning has the advantage that the bandwidth of the video signal can be halved. This is because for each picture scan, the electron beam only has to traverse the screen by one-half the total number of lines. Thus for the same scan rate, the scanning beam needs only to move at one-half the speed so the same horizontal resolution can be achieved with half the bandwidth. This means that the *picture frequency* (that is, the rate at which complete pictures are displayed) is one-half the *field frequency* (that is, the rate at which the electron beam scans the picture area from top to bottom).

For various reasons, the field frequency is usually chosen to be the same frequency as the public electricity supply frequency — for example, 50 Hz in Britain, 60 Hz in the USA. This means that, in Britain, the picture frequency is 25 Hz.

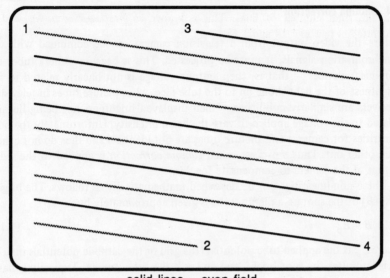

solid lines = even field
broken lines = odd field
(numbers refer to the scan points in figure 1.14)

Figure 1.15 Interlace scans on a TV screen

If 2:1 picture/field interlace scanning can halve the required bandwidth, then the question is can a 3 or 4:1 ratio be used to reduce the bandwidth even further? Unfortunately, the results of these scanning arrangements are not satisfactory since the reduced picture frequency tends to make the picture appear to vibrate up and down.

Given the picture frequency (F_p) and number of lines (N), then it is a simple matter to work out the time interval for each line scan (T_l) using the formula

$$T_l = \frac{1}{NF_p} \tag{1.8}$$

or, assuming a 2:1 interlaced scanning

$$T_l = \frac{2}{NF_f} \tag{1.9}$$

where F_f is the field frequency.

Thus for the 625-line 50 Hz field frequency system as used in Britain:

$$T_l = \frac{2}{625 \times 50} = 64\ \mu s \tag{1.10}$$

Note that this time includes not only the active part of the line scan, but the flyback time as well.

1.4 Gamma correction

Before the video output from a television camera can be combined with the synchronisation signals, it must be processed. This is because camera tubes are non-linear devices — that is, their output voltage is not linearly related to the brightness of the light shone on to the tube face. Also, picture tubes in television receivers are similarly non-linear — their displayed brightness not being linearly related to the voltage applied to vary the beam intensity. Unfortunately the non-linearities for cameras and picture tubes are not identical and thus do not cancel each other out. Thus a process called *gamma correction* is applied to the video output of the camera to compensate for this.

These non-linearities can be described mathematically as follows. The brightness (B) of the spot on a picture tube is given approximately by

$$B \propto E_p^\gamma \tag{1.11}$$

where E_p is the applied tube potential (the grid or the cathode potentials may be varied to alter the spot brightness) and γ is the *tube gamma*. Typically γ has a value between 2 and 3.

Similarly, the video output voltage (E_c) of a camera tube is given approximately by

$$E_c \propto B_s^{\gamma_s} \tag{1.12}$$

where B_s is the scene brightness and γ_s is the *camera gamma* which has a value between 0.3 and 1 depending on the type of camera.

Thus the picture tube brightness is related to the scene brightness by

$$B \propto B_s^{\gamma \gamma_s} \tag{1.13}$$

The exponent term, in this case $\gamma\gamma_s$, is called the *system gamma*. If the system gamma has a value of 1, then the brightness of the displayed picture will be linearly related to the scene brightness.

However, tests have shown that a more satisfactory picture is obtained with a system gamma of around 1.2. This means that the brightness range is in fact compressed, which as well as giving a subjectively better picture also gives a better noise performance since the video amplitude range is reduced.

To obtain the required system gamma, the camera output is passed through a non-linear circuit called a *gamma correction* circuit. Thus the transmitted video signal E_t is related to the scene brightness by

$$E_t \propto B_s^{\gamma_c \gamma_s} \tag{1.14}$$

and the picture tube brightness is related to the scene brightness by

$$B \propto B_s^{\gamma \gamma_c \gamma_s} \tag{1.15}$$

where γ_c is the effect of the gamma correction circuit. The term $\gamma_c \gamma_s$ is usually called the *transmitted gamma*.

Thus to obtain a system gamma of 1.2, γ_c is adjusted such that

$$\gamma\gamma_c\gamma_s = 1.2 \tag{1.16}$$

For a typical value of 2.8 for γ, γ_c therefore would be adjusted so that the transmitted gamma, $\gamma_c\gamma_s$, is $1.2 \div 2.8 = 0.43$.

It is usual to denote the gamma corrected video signal by E', E being the uncorrected signal. This convention will be adopted in this book.

1.5 Synchronisation

So far we have been concerned only with the video information, but just as important is the synchronisation information that must be sent from the camera to the TV so that the scanning motions of the electron beams are kept in step, both in the horizontal and vertical directions. This process involves the transmission of *synchronising pulses* (or *sync pulses*) to control the start of each line and each field scan. These pulses appear during the line and field flyback periods when there is no video information. There are two types of synchronising pulses — line syncs and field syncs.

Line syncs are short pulses that indicate the start of each line, actually the start of each line flyback period as it will be shown in more detail later. Because the horizontal line scanning of the screen never stops, even during field flyback, these pulses are always present.

The *field sync* pulses are more complicated than the line syncs, since they need to incorporate the line syncs as well — as described above; also because of interlace, the start of even fields (that is, fields with even line numbers) and the end of odd fields have to occur half way through a line scan (see figure 1.14). The field syncs consist basically of a set of much wider line sync pulses which will be studied in more detail in the next section.

From these two types of pulses both the line and field scanning mechanisms in the TV set can be kept locked to the scanning in the camera. Usually, however, at the transmission end there is a master sync generator to which all cameras, monitors and other equipment are locked.

1.6 The composite video signal

Since the synchronisation pulses only occur during line or field flyback when there is no video information, these pulses can simply be added to the gamma corrected video signal to produce a single signal which contains all the information necessary to display a fully locked television picture.

This combined signal is called the *composite video signal*. The sync pulses are combined in such a way that they can easily be separated from the video signal in the television. Figure 1.16 shows the composite video signal for two line scans, and figure 1.17 is an oscilloscope photograph of an actual line scan in the 625-line system.

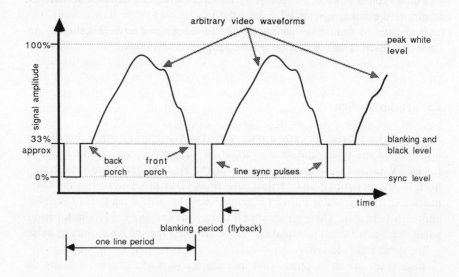

Figure 1.16 Composite video waveform at line rate

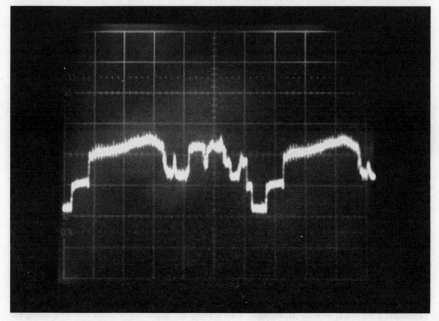

Figure 1.17 Oscilloscope photograph of a line scan in the 625-line system

The ratio between the sizes of the sync pulses and the allowable range of video signal from black to white is carefully chosen to ensure that in the presence of noise (that is, a weak signal) the picture does not become unwatchable before synchronisation is lost, and vice versa. Thus ideally, synchronisation should be lost just as the picture gets so noisy that it is impossible to view satisfactorily. In practice a ratio of around 1:2 for the sync-video signals respectively is chosen.

The various sync and video levels are given names as follows. The maximum video level for white is the *peak white level*, the minimum video level for black is the *black level*, the level below which the display should be switched off (*blanked*) is the *blanking level*, and the level of the sync pulses is called the *sync level*. In most systems, the black level is at the same level as the blanking level (as shown in figure 1.16), but this is not always the case.

The timing datum for the composite signal is the leading edge of the line sync pulse. This leading edge, when differentiated, produces a short pulse which is used to control the start of flyback in the television's line scan generator. This edge is preceded by a short interval where there is no video (that is, black level) which allows the video signal to return to black level, so as not to upset the timing of the line flyback. This is known as the *front porch*. It is important that line flyback occurs at precisely the same point on each line if the resulting picture is not to have a ragged left edge. The front porch ensures that this happens whatever video information occurred at the end of the previous line.

After the line sync pulse is a longer section without video which is called the *back porch*. This allows extra time for the line flyback to be completed, and ensures that the electron beam is blanked throughout this period.

Below are the timings for the various parts of the line sync pulse in the 625-line system.

Front porch: $1.55\ \mu s \pm 0.25\ \mu s$
Line sync pulse: $4.7\ \mu s \pm 0.2\ \mu s$
Back porch: $5.8\ \mu s \pm 0.6\ \mu s$
Overall line sync period: $12.05\ \mu s \pm 0.25\ \mu s$ ($12\ \mu s$ nominal)

There are potentially two methods of differentiating between the line and field syncs — by pulse height or pulse width. Pulse height is not practical because the field sync pulses cannot have amplitudes greater than the line syncs or they would be confused with video information. If they were at a lower amplitude, in the presence of a weak or noisy signal field synchronisation would be lost before line synchronisation. Thus the overall performance of the system would be degraded. Hence differentiation is achieved by pulse width, and the field syncs consist of a set of wider line sync pulses. Either side of these pulses are a set of pulses at twice the line frequency, known as *equalising pulses*.

Electronic integrating circuits are used to detect the wide pulses (known as the *field broad pulses*) and this indicates to the field scan generator the start of the field flyback. The equalising pulses ensure that there is a similar pulse pattern immediately before and after the field broad pulses for both even and odd fields. This assists the field pulse detection circuitry so that the field scans are correctly interlaced. This is critical, since a shift of about 5 per cent of a line period between even and odd fields causes noticeable bunching of pairs of lines on the screen.

Figure 1.18 shows a diagram of the field blanking periods in the composite waveform for even and odd fields (625-line system). Notice that there is always a line sync pulse to keep the line scan generator locked (the extra half line pulses during the equalising and field broad pulses are ignored by the line scan circuitry).

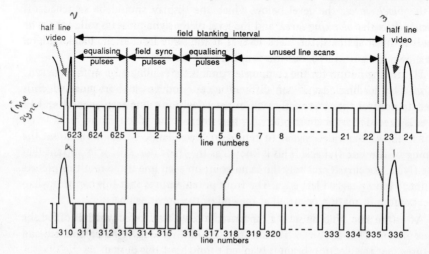

Figure 1.18   Field flyback intervals for the 625-line system

For the field sync interval, the following specifications are set for the various parameters in the British 625-line system.

Number of equalising pulses: 5
Number of field broad pulses: 5
Field broad pulse width: 27.1 μs
Equalising pulse width: 2.3 μs

The start of the field is defined to be at the leading edge of the first field broad pulse.

Thus, to summarise, the composite video signal is made up of two components: the video information which occupies about two-thirds of the total available amplitude, and the synchronisation signals which occupy the remaining one-third. The synchronisation signals consist of line and field pulses, where the field synchronisation consists of a number of broad line sync pulses surrounded by a set of equalising pulses at half line rate. The synchronisation data points are the leading edge of the line sync pulse for line flyback and the leading edge of the first broad pulse for the field flyback.

It must be remembered that the form of the signal described above was developed to be easily decoded by analogue circuitry. In many of today's TVs, digital circuitry is employed (using counters to count line numbers, and timers to detect the field broad pulses, etc.), and therefore if such a system was being designed now, it would be made much simpler (the equalising pulses would not be necessary, for example).

1.6.1 Bandwidth of the composite video signal

Having discussed the form of the composite video signal, it is important that its bandwidth should be determined so that a suitable transmission channel can be used.

At the low-frequency end, a picture containing a uniform brightness (that is, all white or all black or all one shade of grey) contains video frequencies down to d.c. However, the presence of the sync pulses ensures that there must always be a non-zero frequency component at line rate.

At the high-frequency end, the required frequency is determined by the amount of horizontal resolution that is required in the picture. To determine this, it is reasonable to make the horizontal resolution comparable with the vertical resolution, which, as we have seen, depends on the number of lines in the picture.

We have seen that in order to maintain a vertical resolution of 425 lines (determined by the visual acuity of the eye and the average viewing distance) we need to increase the actual number of lines used by the inverse of the Kell factor (K), which is about 0.7. Or, to put it the other way round, for a system with N_a scan lines, then the resulting vertical resolution will be

$$N_a K \text{ lines} \tag{1.17}$$

The value for N_a, however, is the number of active lines that form the picture. For a system with N total scan lines, some will occur during field flyback, and the number of active lines therefore is given by

$$N_a = N a_f \tag{1.18}$$

where a_f is a factor called the *active field factor*.

Hence, for a system with N total line scans, the actual vertical resolution is

$$N a_f K \text{ lines} \tag{1.19}$$

Thus a system with N scan lines should be able to resolve a picture containing the above number of horizontal equally spaced bars. If the horizontal resolution is to be the same as this, then the bandwidth must be sufficient to resolve a picture containing vertical bars of the same spacing. However, the TV picture is not square and usually the picture is wider than it is high. Hence, there would need to be $N a_f K A$ equally spaced vertical bars to give the same bar spacing as $N a_f K$ horizontal bars, where A is the *aspect ratio* of the picture.

Once again, however, this number of vertical bars must be displayed during the active part of a line scan which is a time $T_1 a_1$, where T_1 is the time for one complete line scan (including flyback) and a_1 is the *active line factor*.

In electronic terms, a picture of equally spaced vertical bars corresponds to a square wave video signal, as shown in figure 1.19.

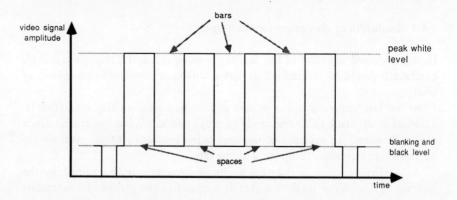

Figure 1.19 Line rate video waveform for three vertical bars

If there are $Na_f KA$ vertical bars, then the frequency of this square wave will be

$$Na_f KA \times \frac{1}{2} \times \frac{1}{T_1 a_1} \qquad (1.20)$$

For a television system with 2:1 interlace, N lines and field frequency F_f:

$$T_1 = \frac{2}{NF_f} \qquad (1.21)$$

hence the frequency of the square wave is

$$f = Na_f KA \times \frac{1}{2} \times \frac{NF_f}{2a_1} \qquad (1.22)$$

$$= \frac{N^2 a_f KAF_f}{4a_1} \text{ Hz} \qquad (1.23)$$

(note that the Kell factor is not involved again in the horizontal resolution calculations since the horizontal signal is continuous).

Fourier analysis shows that the frequency components of this square wave signal contain frequencies of f, $3f$, $5f$ . . . etc., however, a reasonable approximation to a set of vertical bars can be obtained if only the fundamental is transmitted (that is, the bars are represented by a sinusoidal signal rather than a square wave).

Thus the formula for f in equation (1.23) above gives a value for the maximum upper frequency of the video waveform if the horizontal and vertical resolutions are to be the same.

For the British 625-line system:

N = 625 lines
a_f = 0.922
K = 0.7
A = 4/3
F_f = 50 Hz
a_1 = 0.812

which gives

f = 5.175 MHz (1.24)

Comparing this with the bandwidth of, say, 20 kHz for an audio signal, it can be seen how much greater the bandwidth of a video signal is.

1.7 Transmitting the video signal

Before the final composite video signal can be transmitted, the sound information must be added. Clearly this cannot be directly added to the video signal since this signal already contains frequency components in the audio range. The sound therefore is modulated on to a carrier which is outside the video spectrum. Actual methods of modulation and carrier frequencies vary from system to system, but in the British 625-line system, the sound is frequency modulated on to a 6 MHz carrier with a maximum deviation of ± 50 kHz. This method of modulation allows the transmission of high-quality, interference free sound. As with FM radio broadcasts in the VHF band, pre-emphasis (increasing the relative amplitude for higher frequencies) is added to the signal to improve the noise performance. The frequency components of the modulated audio signal are thus placed just above the highest video frequency.

This combined signal may now be modulated on to a radio frequency carrier for broadcast over the air. In fact, for transmission purposes the sound may be frequency modulated directly on to a radio frequency carrier which (for the British system) is 6 MHz above the carrier used for the composite video signal. The effect is the same. Also, some systems use negative modulation for the video signal where peak white corresponds to a low carrier level and sync level corresponds to the maximum carrier level.

Television services are broadcast in the VHF (Very High Frequency) and UHF (Ultra High Frequency) bands and all systems use amplitude modulation for the composite video signal. Single sideband amplitude modulation is the most efficient method in terms of bandwidth and transmitter power, but this method of modulation requires complicated demodulation equipment in the receiver. Hence a compromise is reached, which, at the expense of a small amount of extra bandwidth, allows a relatively simple envelope detector to be used in the television receiver. The method of modulation used is called *vestigial sideband* modulation which contains one complete sideband plus a small section of the other sideband as shown in figure 1.20.

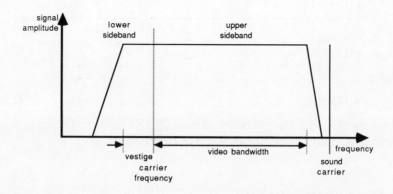

Figure 1.20 Vestigial sideband modulation

Vestigial sideband modulation allows simple envelope detectors to be used in the receiver, without introducing the distortion that usually occurs when demodulating signals that have portions of their spectrums removed. The 'vestige' can be either from the upper or lower sideband and different systems use both methods. The amount of vestige introduced also varies between systems, common values are 1.25 MHz or 0.75 MHz for the bandwidths of the vestige signal. This method is thus universally used for television broadcasts. Figure 1.20 also shows the position of the sound carrier which may be modulated together with the video signal, or by a separate transmitter as mentioned above.

Figure 1.21 shows the complete characteristics for one UHF television channel for the British 625-line system. Notice the *guard band* which ensures adjacent channels do not interfere. UHF transmissions are in band IV (470-610 MHz) and band V (610-940 MHz), and each channel is identified by a channel number from 21 to 68 with increasing frequency.

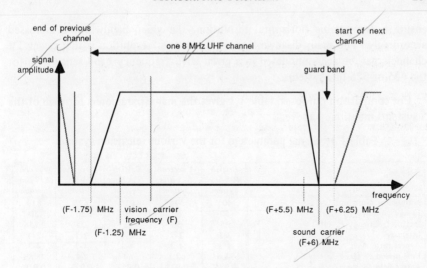

Figure 1.21 British 625-line UHF channel characteristics

1.8 System variations

Unfortunately, there is no universal system for the method of forming the video signal and its modulation on to a radio frequency carrier. There are five main alternative systems in use in the world today. These are:

(a) The British 625-line system
This is used in Britain for all television broadcasts in the UHF band.
(b) The British 405-line system
This was the first system to be used in Britain for television signals in the VHF bands. It has recently been phased out in favour of the superior quality 625-line system above.
(c) European 625-line system
This system is basically identical to the British 625-line system and is used in many European countries. The main difference is in a slightly reduced video bandwidth and different sound carrier frequency (5.5 MHz).
(d) American 525-line system
This is the system used in USA, Canada, Japan and Mexico. The main difference between this system and the 625-line system is the use of a 60 Hz field frequency (corresponding to the 60 Hz electricity supply). This results in many of the other parameters being slightly different. The resolution is also slightly inferior to the 625-line systems.
(e) French 819-line system
This is the system used mainly in France, and, as its name suggests, the picture is divided into 819 line scans. This results in a superior vertical resolution and to

ensure a comparable horizontal resolution, the video bandwidth is increased accordingly. The main disadvantage of this higher resolution is that fewer TV channels can be accommodated in a given UHF frequency band range than for the 625 or 525-line systems.

For comparison purposes, table 1.1 gives the major parameters for each of the 5 systems mentioned above.

Table 1.1 Signal parameters for the various television systems

	British 405-line system	American 525-line system	British 625-line system	European 625-line system	French 819-line system
Number of scan lines (N)	405	525	625	625	819
Field frequency F_f (Hz)	50	60	50	50	50
Picture frequency F_p (Hz)	25	30	25	25	25
Interlace	2-1	2-1	2-1	2-1	2-1
Aspect ratio A	4-3	4-3	4-3	4-3	4-3
Line period T_l (μs)	98.8	63.5	64	64	48.84
Active field factor a_f	0.931	0.923	0.922	0.922	0.919
Active line factor a_l	0.814	0.826	0.812	0.812	0.805
Vertical resolution (lines)	264	340	402	402	527
Horizontal resiolution (lines)	483	422	572	520	786
Carrier modulation	+	−	−	−	+
Black level (% carrier)	35	75	77	75	25
Blanking level (% carrier)	30	75	77	75	25
Peak white (% carrier)	100	15	20	10	100
R.f. channel bandwidth (MHz)	5	6	8	7	14
Vestigial sideband	upper	lower	lower	lower	upper
Vestigial sideband bandwidth (MHz)	0.75	0.75	1.25	1.25	2
Sound modulation	a.m	f.m	f.m	f.m	a.m
Sound carrier w.r.t. vision carrier (MHz)	−3	+4.5	+6	+5.5	−11.15
Sound carrier deviation (kHz)		±25	±50	±50	
Sound pre-emphasis (μs)		75	50	50	

1.9 Summary

The picture is divided into horizontal strips by being scanned with an electron beam. Horizontal scans are called *line scans* and vertical scans are called *field scans*. The pattern of scans is called a *raster*.

The vertical resolution of the picture is determined by the number of lines in the picture, and the rate of scans determines the amount of apparent flicker present. Video bandwidth can be reduced by *interlacing* the scans whereby alternate lines are scanned on one scan and the intermediate lines scanned on the next scan.

To correct for non-linearities in television camera and picture tubes, gamma correction is applied to the video output from the camera.

To ensure synchronisation of the scanning mechanisms between the camera and television, line sync and field sync pulses are added to the video signal during the electron beam flyback periods when there is no picture information.

The ratio of the amplitude of the sync pulses and video information in the composite video signal is chosen for best noise performance.

The bandwidth of the video signal is determined by considerations of the required horizontal resolution and is usually arranged to be about the same as the vertical resolution determined by the number of scan lines.

The sound signal is modulated on to a separate carrier so as not to interfere with the video frequency components.

The composite video signal is amplitude modulated on to a radio frequency carrier using vestigial sideband modulation which allows simple envelope demodulation in the television receiver at the expense of a small amount of extra bandwidth.

2 Colour Television

This chapter will look at how a colour picture can be converted into electrical signals, and how the composite monochrome video signal can be modified to produce a compatible and reverse compatible colour composite signal.

The three principal methods of encoding the colour information — NTSC, PAL and SECAM — will be described.

2.1 The three colour signals

Chapter 1 has explained how any colour can be produced by a suitable combination of red (R), green (G) and blue (B) — the three primary colours.

At this stage it is important to differentiate between adding pigment or paint and adding different coloured lights. Schoolchildren discover that they can make green paint by mixing blue and yellow paints together; similarly, purple paint can be made by mixing red and blue paint, and so on.

The pigments (paints, dyes etc.) appear to be their own colour because they absorb all the others. A red object, seen under a white light, appears red because it reflects red and absorbs all other colours in the white light. If blue light is shone on to a red object there is no red light to reflect, therefore the object appears black.

A mixture of pigments of all colours illuminated by white light appears black because all the colours in the white light have been absorbed. In practice, however, paints and dyes reflect light over a greater range than their nominal colour. Therefore, when yellow and blue pigments are mixed, the resulting colour is green, since this is the colour common to both. A mixture of pigments is said to form a subtractive mixture because the mixture absorbs all colour except those common to each one.

Light, on the other hand, forms additive mixtures. If two coloured lights are shone together on to a white screen, the screen appears to be the colour of their sum — for example, red and green gives yellow.

As explained in chapter 1, the light from a coloured scene can be analysed into combinations of the three primary colours of light — red, green and blue. By recombining these three colour components, the original coloured scene will be reproduced. A colour television service could therefore be provided by using a camera to produce three signals corresponding to the red, green and blue components of the scene. These signals could be transmitted and recombined either optically or electronically in the display device to reproduce the original scene as shown in figure 2.1. Such a system would give excellent pictures but would suffer from the fact that it would not meet the requirements of a commercial colour television system — that is, it is not compatible with the already established monochrome system.

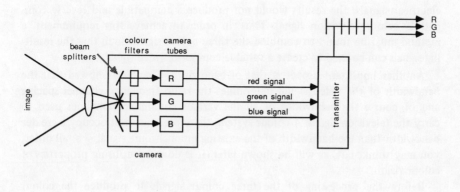

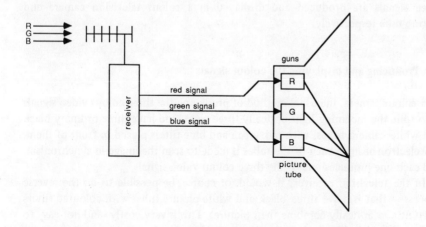

Figure 2.1 Colour transmission using the three primary colour signals

2.2 Compatibility

One of the most important aspects of upgrading any system is that the new system should be compatible and reverse compatible.

In terms of the composite television signal, this means that to retain compatibility an existing monochrome television receiver must be able to display a black and white image when fed with a composite video signal that includes the colour information (that is, the colour composite signal must be *compatible*) and also a colour television receiver must be able to display a black and white picture when fed with a monochrome composite signal as described in chapter 1 (that is, the signal must also be *reverse compatible*).

If the three colour signals were transmitted unaltered as three separate electrical signals, the results would not produce a compatible and reverse compatible colour television signal. Thus, in order to achieve this requirement, a method must be found to combine the three colour signals such that the resulting signals can be used to create a suitable composite video signal.

Another important aspect is that of bandwidth. Chapter 1 showed that the bandwidth of the video signal determines the radio frequency channel spacing and of course the bandwidth of all the various communication lines used to carry the television signal. A colour system, therefore, must not occupy a greater bandwidth than the bandwidth of the existing monochrome signal — a tall order, you may think, but, as will be shown later, it is done by utilising properties of colour vision.

Before the processing of the three colour signals to produce the colour composite video signal is discussed, it is helpful to look briefly at how these three signals are produced and displayed in a colour television camera and picture tube respectively.

2.3 Producing and displaying the colour signals

In a colour camera, the usual method of producing the three colour video signals is to split the incoming light optically three ways, and have three ordinary black and white camera tubes, with red, green and blue filters placed in front of them. The electron beam in each of the tubes is made to scan the image in synchronism, and each one produces one of the three colour video signals.

In the television receiver, it would, of course, be possible to do the reverse process — that is, have three black and white picture tubes with coloured filters in front and optically combine their pictures. This is very costly, and not easy to achieve. Instead, the usual method is to combine three electron beams into one picture tube, as described below.

2.3.1 *Colour television tubes*

On the screen of the colour television tube is a set of three coloured phosphors (which produce coloured light when hit by an electron beam) arranged in some pattern. There are different details here, but one of the most popular methods is to have the phosphors arranged in sets of triangular dots as shown in figure 2.2.

Three electron beams are incorporated into the picture tube, and each of these is only allowed to hit one colour of phosphor dots by the use of a perforated metal mask located just behind the screen. This is called the *shadow mask*. The sets of dots are made to be small enough so that at normal viewing distances the individual dots are indistinguishable, and also the relative brightness of the colours in each set of three dots merges to produce the required actual colour of the picture at that point.

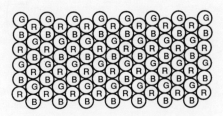

Figure 2.2   Phosphor dots on a colour television tube

The three electron beams are made to scan the screen in synchronism and the three colour signals are arranged to alter the intensity of each of the beams and thus the brightness of the corresponding colour produced on the screen. Hence a full colour picture is perceived when looking at the screen from the normal viewing distance. Figure 2.3 shows a simplified diagram of a colour television picture tube.

2.4 Colour difference signals

Any colour has three characteristics — brightness, hue and saturation. *Brightness* can be readily understood as the intensity of the light, produced or reflected — that is, the black and white component. *Hue* is the sensation which enables us to distinguish between the various colours — that is, to know whether it is red, green or blue etc. *Saturation* is a measure of the purity of the colour — that is, the degree to which the colour is diluted with white light. For example, pink and red have the same hue — pink is merely red light diluted with white. Pink is therefore less saturated than red. Note also that as well as having the same hue, a bright red and a dull pink might also have the same brightness.

There is thus another way of transmitting a colour signal. Instead of the three primary colours red, green and blue, the signals corresponding to the brightness, hue and saturation of the scene could be transmitted. In either case, we need to transmit three signals to enable the full colour picture to be reproduced.

A fact which enables the bandwidth requirements for a compatible colour television system to be met is that the human eye, while being sensitive to

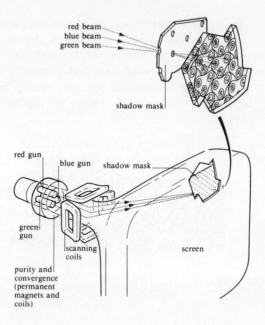

Figure 2.3 Colour television picture tube

changes in brightness, is fairly insensitive to changes in hue. This means that, although colours might diffuse into each other at a transition, the eye sees a sudden change from one colour to the other. In the television sense this means that a wide bandwidth is required for brightness changes while only a narrow band is required for colour (hue) changes. Also if a monochrome receiver is to reproduce a black and white picture from the colour signal, the brightness signal must be created and transmitted as the main video signal within the composite signal. The colour information must then be incorporated into the video signal in such a way that it does not interfere with the operation of a monochrome television.

It has been found that by adding a certain proportion of red (R), green (G) and blue (B) light, then white light can be produced. The actual proportions, given below, produce an equation that produces a brightness signal (Y) (that is, the black and white signal) from the three colour signals:

$$Y = 0.30R + 0.59G + 0.11B \qquad (2.1)$$

The other requirements for hue and saturation are obtained by subtracting Y from both sides of equation (2.1) giving

$$0 = 0.30(R - Y) + 0.59(G - Y) + 0.11(B - Y) \qquad (2.2)$$

For a black and white scene $R = G = B = 1$ giving $Y = 1$ and thus

$$(B - Y) = 1 - 1 = 0 \tag{2.3}$$

$$(R - Y) = 1 - 1 = 0 \tag{2.4}$$

The coefficients in equation (2.2) are called *colour difference signals* or *chrominance signals* since they contain colour information only and no brightness information.

It was shown above that three signals need to be transmitted. If one is to be the brightness (Y) signal, and if two of the three colour difference signals are transmitted as well, then the original three R, G and B signals can be re-created in the receiver. This is so since if, say, the $(R - Y)$ and $(B - Y)$ signals are transmitted along with the Y signal, then the R, G and B signals can be obtained as:

$$(R - Y) + Y = R \tag{2.5}$$

$$(B - Y) + Y = B \tag{2.6}$$

and from equation (2.2):

$$(G - Y) = -\frac{0.30(R - Y)}{0.59} - \frac{0.11(B - Y)}{0.59} \tag{2.7}$$

giving

$$G = Y - \frac{0.30(R - Y)}{0.59} - \frac{0.11(B - Y)}{0.59} \tag{2.8}$$

the third colour signal.

This has the advantage that for all parts of the scene containing no colour, the colour difference signals disappear — see equations (2.3) and (2.4). Also, since the eye is relatively insensitive to colour changes, the chrominance signals which contain hue and saturation information only are narrow band signals compared with the brightness signal.

Thus the brightness signal (Y), known as the *luminance*, can be transmitted as the main video signal in the composite video waveform, and a method must be found of adding two of the three colour difference signals to this signal in such a way as not to affect the operation of black and white receivers.

It is at this point that differences occur between the three principal systems in use today for terrestrial broadcasts. The features of each system will now be considered.

2.5 The NTSC signal

NTSC stands for National Television Systems Committee of America. The NTSC system was developed in the USA and officially adopted there at the end of

1953. This system of encoding the colour video signals is used mainly with 525-line 60 Hz systems in the USA, Canada, Japan and Mexico, although the system is also specified for 405 and 625-line systems as well.

The NTSC composite video signal is made up as explained in the following sections.

2.5.1 Monochrome signal

The main monochrome video signal is obtained from the luminance signal formed from the three colour signals according to the proportions given in equation (2.1) as explained above, thus ensuring monochrome compatibility.

Gamma correction is applied to the camera R, G and B outputs before the luminance signal is formed. This represents a slight loss of compatibility for a monochrome receiver tuned to a colour transmission, since the actual correct monochrome signal is a gamma corrected version of the output of a black and white camera tube, which is equivalent to gamma-correcting the Y signal after it is generated rather than correcting the three colour signals before generation. This error occurs because generation of the gamma corrected signal involves non-linear functions (that is, $R^\gamma + G^\gamma + B^\gamma$ is not the same as $(R + G + B)^\gamma$).

This system of gamma correction is used to simplify the receiver circuitry in a colour television, since if the correct method was used, non-linear networks would be needed in the receiver to regenerate the green colour signal. The fact that the colour difference signals are band-limited also means that there is a slight gamma error on a colour picture for frequencies above the colour difference signal frequency range.

Many tests have shown that the resulting degradation in performance due to these gamma errors is far outweighed by the simplification in receiver design achieved.

2.5.2 Chrominance signals

As mentioned in the previous section, three components must be contained in the video signal in order to be able to reconstitute the original R, G and B signals in the receiver.

For compatibility, one of these signals must be the Y' luminance signal. The others must therefore be two of the three colour difference signals and the $(R' - Y')$ and $(B' - Y')$ gamma corrected signals are used.

How are these signals added to the luminance signal? They cannot be added directly since their bandwidths coincide with that of the Y' signal. Instead, the colour difference signals are modulated on to a carrier frequency near the top of the video band. This is possible because, as already explained, because of the eye's relative insensitivity to colour changes, the two colour difference signals can have a reduced bandwidth, and thus only a small band of frequencies at the top of the luminance band need be occupied by the modulated colour difference signal.

The chrominance signals are quadrature-modulated on to a carrier called the *colour sub-carrier* which has a frequency near the top of the video band. Quadrature modulation allows the two signals to be separately modulated on to the same carrier. In phasor terms, the blue difference signal phasor is usually used as the reference (ref.) phase, and the red difference signal leads this by 90°, as shown in figure 2.4. The resulting modulated chrominance signal is called the *chroma signal*.

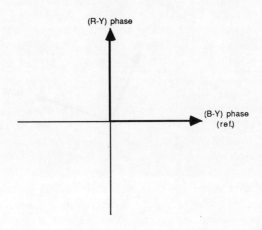

Figure 2.4 Chrominance signal phasors

A chrominance bandwidth of around 1 MHz is required for acceptable colour pictures but, if double sideband modulation were used, this would occupy nearly half the luminance bandwidth in the 525-line system. Studies have shown however that the eye is more sensitive to fine colour changes in the orange/cyan than in the green/magenta colour ranges. This means that the modulation of the colour difference signals can be achieved with less bandwidth, as follows.

First, in order not to overload the transmitter with the modulated colour difference signals when they are added to the luminance signal, they are reduced in amplitude and are usually termed the U and V signals as below.

$$U = 0.493(B' - Y') \tag{2.9}$$

$$V = 0.877(R' - Y') \tag{2.10}$$

These signals are called the *weighted colour difference signals*.

Second, in order to be able to minimise the bandwidths of one of the quadrature modulated signals, from considerations of the eye's sensitivity to fine colour changes these U and V signals are not modulated directly on to the blue and red sub-carrier phases. Instead the chrominance subcarrier phases are advanced by 33° from the $(B - Y)$ and $(R - Y)$ phases mentioned above and are modulated by two further signals, I and Q, which have components of the U and V signals as below.

$$I = V \cos 33° - U \sin 33° \qquad (2.11)$$

$$Q = V \sin 33° + U \cos 33° \qquad (2.12)$$

This is shown on the phasor diagram of figure 2.5.

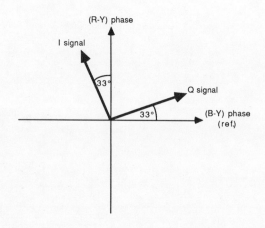

Figure 2.5 I and Q signal phasors

In fact, the U and V signals need not be generated, since the I and Q signals can be obtained directly by combining equations (2.9) to (2.12) to give the following result:

$$I = 0.74(R' - Y') - 0.27(B' - Y') \qquad (2.13)$$

$$Q = 0.48(R' - Y') + 0.41(B' - Y') \qquad (2.14)$$

The result of producing these two signals means that the bandwidth of the Q signal can be reduced below that of the I signal without significantly affecting the resulting visual appearance of the picture.

These two signals, then, are quadrature-modulated on to the sub-carrier with phases of (ref. + 33)° and (ref. + (90 + 33))° respectively for the Q and I signals, and added to the monochrome luminance signal. In order to reduce the bandwidth of the modulated signal as much as possible, the larger bandwidth I signal is modulated using vestigial sideband modulation, while the Q signal uses full double sideband modulation. The vestige used for the I signal is the upper side-

Colour Television

band and its bandwidth coincides with the Q signal upper sideband bandwidth. Both signals use suppressed carrier modulation.

The actual values have been specified for 405, 625 and 525-line systems, although the NTSC system is only usually used with 525-line systems. Table 2.1 shows the relevant parameters for each of these systems, and figure 2.6 shows the frequency spectrum of the NTSC composite video signal for a 525-line 60 Hz system.

Table 2.1 NTSC parameters for different line systems

Line System	Luminance bandwidth MHz	I bandwidth MHz	Q bandwidth MHz
405	3.0	0.5	0.3
525	4.2	1.3	0.5
625	5.5	1.6	0.8

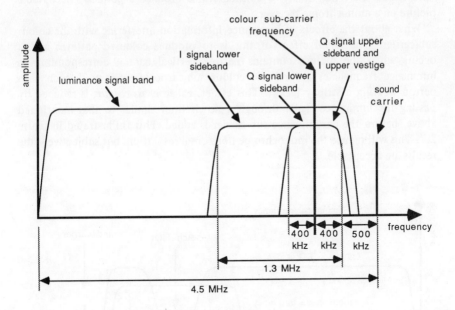

Figure 2.6 NTSC signal spectrum

2.5.3 Luminance/chrominance interference effects

The NTSC composite video waveform then is composed of the addition of the monochrome luminance signal, the modulated I and Q chrominance signals and, of course, the sync pulses and sound sub-carrier as in the monochrome system.

What is the effect on the luminance signal from interference by the chrominance sub-carrier, and vice versa, since the chrominance sub-carrier lies at the top of the luminance band?

The former effect can be seen as a series of spots across and down the screen. The brightness of these spots depends on the saturation of the picture at that particular point (since the amplitude of the sub-carrier depends on the amplitude of the chrominance signals which determines the colour saturation). Subjectively, these spots can be made least noticeable by a combination of three methods.

First the amplitude of the sub-carrier is made as small as possible, consistent with an acceptable noise performance. Second the frequency of the sub-carrier is carefully chosen to be in a fixed relation with the line frequency so as to make the spots move down the screen in the least obtrusive way. This aspect will be covered in more detail later. Finally, the luminance section of the receiver contains a 'notch' filter which attenuates frequencies in the sub-carrier region. Such a filter would not be present in an older monochrome receiver, but experiments have shown that the first two considerations produce a perfectly acceptable picture on a monochrome receiver.

What about the effects of luminance information interfering with the colour sub-carrier? The visual effect of this is to produce coloured patterns and it occurs when the picture contains narrow vertical bars — corresponding to luminance frequencies around the colour sub-carrier frequency — such as a person wearing a stripey jacket. This effect, called *cross-colour*, is reduced by passing the luminance signal through a notch filter similar to that mentioned above, before the chrominance sub-carrier is added. This is illustrated in figure 2.7. This will reduce the monochrome horizontal resolution, but subjectively the results are acceptable.

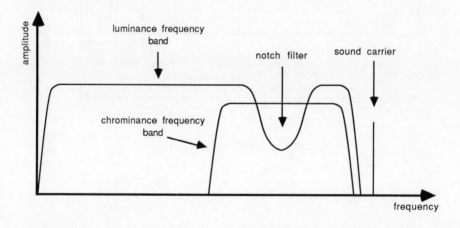

Figure 2.7 Luminance notch filter

Colour Television

It may be thought that this method of encoding colour pictures is a bit messy, but many experiments have shown that the end result is perfectly acceptable, and does meet the criterion of being compatible and reverse compatible. Note also the advantage of using amplitude modulation for the chrominance signals since for any portions of a picture which are in black and white, the sub-carrier signal simply disappears and there are no cross-colour effects.

2.5.4 Sub-carrier burst

There is one final component of the NTSC composite signal that has not been mentioned.

Demodulation of a suppressed carrier modulated signal requires a local in-phase sub-carrier to be available. This cannot be generated from the chroma signal, so an alternative must be used.

The solution is to insert a small 'burst' of unmodulated sub-carrier on every scan line in the back porch after the line sync pulse. This signal, containing only about nine cycles of sub-carrier, is gated out in the receiver and used to phase lock a local sub-carrier oscillator. This oscillator is stable enough to remain sufficiently in phase until the burst at the start of the next scan line. The phase of the burst is 180° out of phase with the U signal reference phase as shown in figure 2.8. Thus in the receiver this locally generated sub-carrier can be used with suitable phase shifting circuits to demodulate both the chrominance signals.

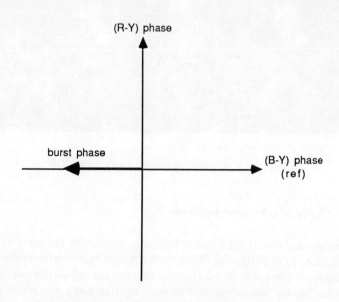

Figure 2.8 NTSC burst phase

38   *Television and Teletext*

This burst signal has another use in that its presence indicates colour and its absence monochrome transmissions (if the broadcaster switches off the burst accordingly). This can cause circuitry in a receiver to switch off its colour decoding circuitry and give a better noise performance for monochrome pictures.

Note that the burst is present at the start of every scan line except those during the field sync and equalising pulses.

Figure 2.9 is an oscilloscope photograph of a line scan from a colour composite video signal, and figure 2.10 is a photograph of an enlarged section of a line sync pulse showing the burst in more detail. Note the varying 'thickness' of the trace in figure 2.9 caused by the varying chroma signal amplitude on top of the luminance signal.

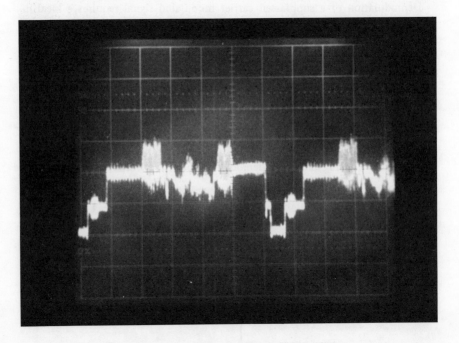

Figure 2.9   Colour composite video waveform

2.5.5 *Choice of sub-carrier frequency*

As explained above, the sub-carrier frequency is carefully chosen to minimise the appearance of dot patterning. What are the factors that determine the choice?

Clearly, if there was no relationship between the sub-carrier and line frequencies (that is, the sub-carrier was allowed to 'free run'), the result would be a constantly changing pattern that would sometimes be static and sometimes move. Thus there is a need for some relationship between the two frequencies. Also if

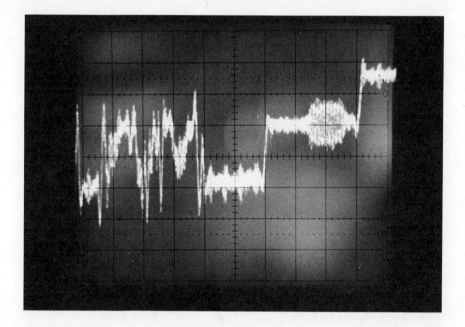

Figure 2.10 NTSC burst signal

there is an exact number of sub-carrier cycles in each line, then there would appear an objectionable static pattern of vertical or slanting bars. Experiments have shown that the interference is least noticeable when the sub-carrier frequency is an odd number integer multiple of half line frequency. That is

$$f_{sc} = (2n + 1) \times \tfrac{1}{2} f_1 \qquad (2.15)$$

where f_{sc} = sub-carrier frequency
f_1 = line frequency
and n is an integer.

This means that since there is an odd number of half cycles of sub-carrier in one line and the number of lines per picture is odd, it will take four fields (two pictures) for the dot pattern to repeat. The pattern thus appears to move slowly down the screen. However the inclusion of changing colour information causes both random phase and amplitude changes in the dot pattern — reducing further the subjective effect of the patterning.

How is the value for n chosen in equation (2.15) above? In order to have as little effect on the luminance signal as possible, the sub-carrier frequency should be as high as possible. However it must not be so high that the upper chrominance sideband interferes with the sound carrier or entails making the TV channel bandwidth larger. Also, for convenience, n should be chosen to enable the line and field frequencies to be fairly easily generated from a master sub-carrier oscillator.

For the 525-line 60 Hz system with a video bandwidth of 4.2 MHz, n is chosen as 277 giving $(2n + 1) = 455$, which has factors of 5, 7 and 13. The sub-carrier frequency is set at 3.579 545 MHz resulting in a line frequency of 15.734 264 kHz and a field frequency of 59.94 Hz. These values are sufficiently close to the 60 Hz/15.750 kHz values for the line/field frequencies in the monochrome system so as not to affect monochrome receivers. The frequencies for the other line systems are given in Table 2.2.

Table 2.2 NTSC sub-carrier frequencies for different line systems

Line system	Line frequency Hz	Field frequency kHz	Sub-carrier frequency MHz
405	50	10.125	2.6578125
525	59.94	15.734264	3.579545
625	50	15.625	4.4296875

Note that the system is not locked to the public mains supply frequency since its frequency is not sufficiently stable to give an acceptable stability for the sub-carrier.

2.5.6 Summary

(a) The full bandwidth monochrome signal is formed by adding the R, G and B colour signals in the proportions given in equation (2.1). This signal is called the luminance signal.

(b) Gamma correction is applied to the R, G and B signals before the luminance signal is generated. This results in a slight loss of compatibility for monochrome receivers and for fine colour detail on colour receivers, but is justified by the simplification of the receiver circuitry.

(c) The colour information is provided by including the $(R' - Y')$ and $(B' - Y')$ colour difference signals in the composite signal.

(d) These signals are modulated on to a sub-carrier frequency near the top of the video band using suppressed-carrier quadrature modulation.

(e) To prevent overloading the transmitter with the combined luminance and colour difference signals, the $(B' - Y')$ and $(R' - Y')$ signals are reduced in amplitude to produce the weighted chrominance signals U and V respectively.

(f) Because of the eye's insensitivity to colour changes, the chrominance signals can be narrow band signals.

(g) In addition, because of the eye's relative insensitivity to fine colour changes in the green/magenta colour range, the bandwidth of the U signal can be further decreased.

(h) To maximise the bandwidth reduction described in (g), the U and V signals are not modulated directly, but two further signals Q and I are produced which are used to quadrature-modulate the colour sub-carrier with phases of (ref. + 33)° and (ref. + (90 + 33))° respectively. The reference phase is defined as that corresponding to the U signal component. Thus the I and Q signals contain components of both the U and V signals.
(i) The bandwidth of the Q signal is then reduced further and is modulated on to the colour sub-carrier using double sideband suppressed carrier modulation.
(j) The I signal is similarly modulated, but vestigial sideband modulation is used, the vestige being on the upper sideband and having a width equal to the bandwidth of the Q signal.
(k) In order to demodulate the colour difference signals, an in-phase carrier is required in the receiver. A local oscillator is kept in phase by the addition of a 'burst' of unmodulated sub-carrier contained in the back porch of every line sync pulse (except pulses occurring during the field sync or equalising pulses). The transmitted phase of the burst is (ref. + 180)°.
(l) The full NTSC composite video signal is therefore composed of the luminance and chroma signals together with the sync pulses, burst and sound carrier.
(m) Interference on the luminance signal from the chroma signal is controlled by reducing the carrier amplitude as much as possible and by carefully choosing its frequency. A frequency relationship of an odd number of multiples of half line frequency is found to produce the best results.
(n) Interference on the chrominance information from the luminance signal is reduced by feeding the luminance signal through a notch filter which is centred on the colour sub-carrier frequency.

At first sight the above method of encoding a colour picture seems unnecessarily complicated and messy. However given the constraints of compatibility and reverse compatibility together with the requirement not to increase the bandwidth of the encoded signal, the result is inevitable. The reproduced picture however is found to be perfectly acceptable and can be achieved without undue circuit complexity in the receiver. Such an achievement is a credit to the designers of the system.

It is important to appreciate how the system maximises the use of the available frequency band and in doing so has to make small compromises to the performance of the original monochrome signal. Also notice how the characteristics of the eye are used to compress the colour information so as to have least effect on the monochrome signal while still allowing an acceptable colour picture to be reproduced.

Figure 2.11 shows a simplified block diagram of an NTSC transmitter. Notice the inclusion of the delay in the I signal to compensate for the extra band limiting of the Q signal, and also the delay in the luminance signal to compensate for the bandwidth reduction of both the chrominance signals. Without these delays, the colour information would not appear at the correct point when overlaid on the monochrome picture.

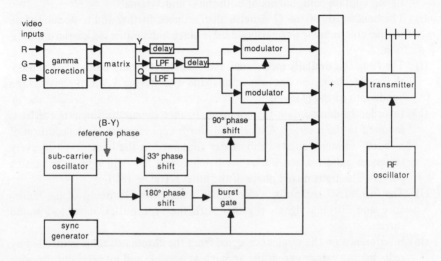

Figure 2.11 Block diagram of an NTSC transmitter

2.6 The PAL system

PAL stands for Phase Alternating Line for reasons which will become clear later. The system was originally developed in Germany in 1962, but was upgraded and modified before being introduced into broadcasts in Britain and Germany in 1967.

2.6.1 Effects of sub-carrier phase errors in the NTSC system

The main disadvantage of the NTSC system which the PAL system attempts to overcome is that of phase errors occurring either in the transmission path or in the receiver itself.

In the NTSC system, the burst is used to synchronise a local colour sub-carrier oscillator to the oscillator at the transmitter in order that the chroma signal can be properly demodulated. What happens, however, if the phase of the colour sub-carrier is disturbed in relation to the burst, either during transmission or in the television receiver?

This effect can occur for a variety of reasons, the main ones being filters and the effects of vestigial sideband modulation of the video signal; such effects are called *differential phase errors*. Let us look at how such errors affect the displayed picture.

Figure 2.12 shows a phasor diagram of a typical colour sub-carrier phasor OA which has components OU and OV of the colour difference signals $(B' - Y')$ and $(R' - Y')$ respectively. Because of differential phase distortion, this phasor when received is demodulated as phasor OA_r. This phasor has $(B' - Y')$ and $(R' - Y')$ components of OU_r and OV_r respectively. Thus on demodulation, the amplitude of the $(B' - Y')$ signal will be less than that transmitted and the $(R' - Y')$ signal will be larger. These colour difference signal amplitude changes will result in changes of hue (colour) on the screen and phase errors of more than ± 5% are noticeable — especially if the errors are not static, resulting in the colours 'wandering'.

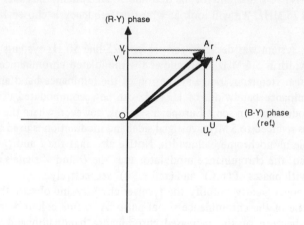

Figure 2.12   Effect of NTSC phase distortion

Other phase errors can occur, resulting in the same visual effect because the colour sub-carrier is riding on top of the luminance signal and thus the sub-carrier's absolute level is not fixed. This can cause circuits in the receiver to produce phase errors called *level dependent phase errors*.

Also, a receiver might have its sub-carrier phase-locked-loop out of adjustment, giving a permanent phase shift in the locally generated sub-carrier. This, however, can sometimes be corrected by adjustment of the front panel 'hue' control — if there is one.

Anyone who has seen an NTSC picture will notice how colours frequently tend to vary — for example, people with faces in odd shades of pink are quite common. Thus it is the effect of these phase errors which is the major disadvantage of the NTSC system and it is this that the PAL system mainly seeks to improve.

There are three main differences in the PAL composite video signal from the NTSC signal. They are in the bandwidths of the colour difference signals, the method of chrominance modulation and the composition of the sub-carrier burst. Let us look at each of these in turn and see how the changes improve the phase error disadvantage of the NTSC system.

2.6.2 Colour difference bandwidths

In the PAL system, the bandwidths of the U and V signals are identical and both are quadrature-modulated using double sideband suppressed carrier modulation. The sub-carrier frequency is chosen from the same sort of considerations as those for the NTSC signal, and for the 625-line 50 Hz system; the exact frequency is 4.433 618 75 MHz. We will look at why this frequency is chosen later in this chapter.

The PAL system was designed for use on 625-line 50 Hz systems where the video bandwidth is 5.5 MHz. This allows the modulated chrominance signals to occupy a larger frequency range at the top of the luminance band and hence a larger chrominance bandwidth of 1.3 MHz can be accommodated using double sideband modulation. In some European systems, however, where the luminance bandwidth is reduced to 5 MHz, vestigial sideband modulation is used to preserve an acceptable monochrome bandwidth. Notice also that the I and Q signals are not generated; the chrominance modulator uses the U and V signals and modulates these with phases of (ref.)° and (ref. + 90)° respectively.

These changes greatly simplify the receiver circuitry and obviate the need for a delay in one of the chrominance signal paths. Also fine colour detail is better reproduced because of the increased chrominance bandwidth and the use of double sideband modulation for both colour difference signals. (Note that in the NTSC and continental PAL systems, the move to single sideband modulation for frequencies above the vestige band causes a halving of the demodulated chrominance amplitude and thus a reduction in saturation for fine colour detail.)

2.6.3 Phase reversal

The second and most important difference from the NTSC system is that the phase of the V sub-carrier is changed by 180° every line. Figure 2.13 shows the phasor diagram of the U and V modulated signals in the PAL system. Notice that there is no 33° offset as with the I and Q signals in the NTSC system.

Why is this phase reversal introduced? The reason is that the hue effects of phase errors mentioned above are averaged over adjacent lines. This can be done either by the eye, or more satisfactorily by electronic circuits as explained below.

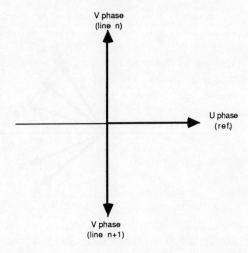

Figure 2.13 PAL chrominance signal phasors

2.6.4 Simple PAL

With *simple PAL*, the PAL decoder in the television uses normal synchronous demodulators with a reversing switch operating at half line frequency on the input of the V demodulator.

When the result is displayed, hue errors resulting from phase errors appear on adjacent lines in opposite directions, and this is perceived by the eye as a mixture which results in the correct original hue. This can be proved simply by considering the phasors involved.

In figure 2.14 there are the following phasors:

OA is the transmitted phasor on line n;
OA_r is the received phasor on line n;
OB is the transmitted phasor on line $(n + 1)$, assuming the same hue and saturation as line n;
OB_r is the received phasor in line $(n + 1)$;
OC_r is the received phasor on line $(n + 1)$ after phase reversal of the V component in the receiver.

The average phasor over the two lines n and $(n + 1)$ can be found by drawing a line between O and the centre of a line between C_r and A_r (OD_r on the figure). This line of course lies along the OA phasor line, but has a slightly shorter length — that is, a slightly lower saturation — but the average hue is the same as that transmitted.

46   Television and Teletext

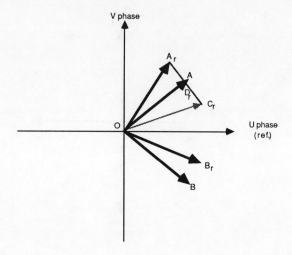

Figure 2.14   Phase distortion in the PAL system

Unfortunately, in practice simple PAL does not work very well for the following reasons:

(a) The averaging process relies on adjacent lines having the same colours — which is not true in practice.
(b) The eye is not a good averager of the hues.
(c) 'Adjacent lines' means adjacent in time — that is, every other line on the TV screen (because of interlace).
(d) As phase errors increase the saturation is reduced.

When looking at a simple PAL picture, the effect of phase errors appears as a set of horizontal moving bars called *Hanover bars*.

A better visual picture can be obtained by performing the averaging process electronically. This involves the use of a delay line and is termed *delay line PAL*.

2.6.5 Delay line PAL

Delay line PAL, which is used in all today's PAL television receivers, involves the use of an exact one scan-line delay.

The modulated chrominance signal is passed through the delay line and the output of this is added and subtracted from the undelayed signal. The outputs of the adder and subtractor are then demodulated using the locally generated sub-carrier to give the unmodulated V and U signals as below:

Colour Television

$$\text{modulated carrier on line } n = (U + V) \tag{2.16}$$
$$\text{modulated carrier on line } (n + 1) = (U - V) \tag{2.17}$$
$$\text{modulated carrier on line } (n + 2) = (U + V) \tag{2.18}$$

Hence:

$$\text{delayed signal during line } (n + 1) = (U + V) \tag{2.19}$$
$$\text{delayed signal during line } (n + 2) = (U - V) \tag{2.20}$$

Therefore on line $(n + 1)$:

$$\text{adder output} = (U + V) + (U - V) = 2U \tag{2.21}$$
$$\text{subtractor output} = (U + V) - (U - V) = 2V \tag{2.22}$$

and on line $(n + 2)$:

$$\text{adder output} = (U - V) + (U + V) = 2U \tag{2.23}$$
$$\text{subtractor output} = (U - V) - (U + V) = -2V \tag{2.24}$$

Thus if the demodulator output is inverted every line (or more usually the sub-carrier input to the demodulator is inverted), the result will produce the separate U and V chrominance signals. This process is illustrated in figure 2.15.

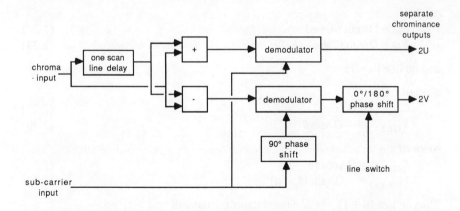

Figure 2.15 Separation of the U and V chrominance signals

What happens now if there are phase errors? Figure 2.16 shows a chroma phasor OA with phase angle θ. On reception the phase angle has changed by an angle α.

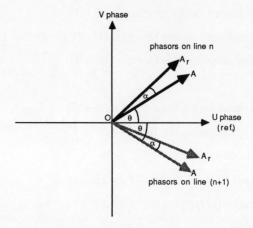

Figure 2.16 Chroma phasor error

The U and V components on line n are as follows:

transmitted
$$U_{tn} = OA \cos \theta \tag{2.25}$$
$$V_{tn} = OA \sin \theta \tag{2.26}$$

received
$$U_{rn} = OA\cos(\theta + \alpha) \tag{2.27}$$
$$V_{rn} = OA\sin(\theta + \alpha) \tag{2.28}$$

and on line $(n + 1)$:

transmitted
$$U_{t(n+1)} = OA\cos \theta \tag{2.29}$$
$$V_{t(n+1)} = -OA\sin \theta \tag{2.30}$$

received
$$U_{r(n+1)} = OA\cos(\theta - \alpha) \tag{2.31}$$
$$V_{r(n+1)} = -OA\sin(\theta - \alpha) \tag{2.32}$$

Thus on line $(n + 1)$, the U demodulator output will be

$$OA(\cos(\theta - \alpha) + \cos(\theta + \alpha)) \tag{2.33}$$
$$= OA((\cos \theta \cos \alpha + \sin \theta \sin \alpha) + (\cos \theta \cos \alpha - \sin \theta \sin \alpha)) \tag{2.34}$$
$$= 2OA\cos \theta \cos \alpha \tag{2.35}$$
$$= 2U_t \cos \alpha \tag{2.36}$$

and the V demodulator output will be

$$\text{OA}\,(-\sin(\theta - \alpha) - \sin(\theta + \alpha)) \tag{2.37}$$

$$= \text{OA}\,(-(\sin\theta\cos\alpha - \cos\theta\sin\alpha) - (\sin\theta\cos\alpha + \cos\theta\sin\alpha)) \tag{2.38}$$

$$= -2\text{OA}\sin\theta\cos\alpha \tag{2.39}$$

$$= -2V_t \cos\alpha \tag{2.40}$$

and similarly for line $(n+2)$ except the V demodulator output will be $+2V_t \cos\alpha$.

Thus a phase angle error of α results in a change in amplitude of both the chrominance signals by the factor $\cos\alpha$ — that is, a saturation change rather than a hue change. (Note that the factor 2 can be catered for in the design of the receiver since it is always present.)

Another advantage of the PAL system is that the U and V sub-carriers are separated before demodulation, allowing the demodulator circuitry to be of a simpler design. Remember though, that complete cancellation of either the U or V signal will only occur when adjacent lines have identical hue and saturation components.

There are two consequences of the PAL system that must be remembered. Firstly because the chrominance components of adjacent lines are 'averaged', the vertical colour resolution is more than halved. (Because of interlace, the averaging is done between alternate lines of the picture.) However, since the horizontal chrominance resolution has been reduced more than this by bandlimiting, this is not such a disadvantage. The second consequence is that phase errors result in changes in colour saturation (rather than hue) which is far more acceptable than the hue changes resulting in the NTSC system.

There is one consideration we have not yet looked at. How does the receiver know whether the current scan line is a line 'n' or a line '$(n + 1)$' so that it can set its 'V switch' correctly? This is done by switching the burst phase as explained below.

2.6.6 Swinging burst

In order for the television receiver to detect the phase of the V sub-carrier signal on each line, the burst phase is also changed on successive lines. The phases chosen are 135° for positive V lines and 225° for negative V lines — that is, 180° − 45° on lines with no V inversion (known as N lines (as in the NTSC system)) and 180° + 45° lines with V inverted (known as P lines (for PAL)). This ± 45° swing is called the *swinging burst* and it allows the television receiver simply to detect whether a P or N line is being transmitted and thus set its 'V switch' accordingly.

2.6.7 Sub-carrier frequency

The phase reversal of the V signal means that the sub-carrier frequency used in the NTSC system is not suitable. The phase switching produces an effect due to the V component as if a sub-carrier frequency of a multiple of line frequency were used rather than an odd multiple of half line frequency as in the NTSC system.

The answer is to use the NTSC value with a quarter line period offset together with an overall picture frequency offset of 25 Hz which additionally produces a field interlace for the resulting dot pattern. This reduces the sub-carrier visibility even further.

The actual frequency is given by the formula:

$$f_{sc} = \tfrac{1}{2}(2n + 1)f_1 + \tfrac{1}{4}f_1 + 25 \text{ Hz} \tag{2.41}$$

For a 625-line system, the NTSC sub-carrier frequency is specified as 4.429 687 5 MHz (that is, $n = 283$) hence the PAL sub-carrier frequency is

$$f_{sc} = \tfrac{1}{2}(2 \times 283 + 1)f_1 + \tfrac{1}{4}f_1 + 25 \text{ Hz} \tag{2.42}$$

$$= 4.433\,618\,75 \text{ MHz } (f_1 = 15.625 \text{ kHz}) \tag{2.43}$$

2.6.8 Summary

To summarise, below are listed the features of the PAL system that differ from the NTSC system:

(a) The original gamma corrected and weighted U and V signals are quadrature-modulated.
(b) Double sideband suppressed carrier modulation is used for both chrominance signals.
(c) The phase of the V sub-carrier is inverted every line.
(d) The phase of the burst is changed by $\pm 45°$ on successive lines, a negative change for N lines and a positive change for P lines.
(e) The colour sub-carrier has a quarter line offset to reduce patterning and a 25 Hz offset to reduce sub-carrier visibility. The frequency chosen is 4.433 618 75 MHz, which is given by equation (2.41).

Figure 2.17 shows a block diagram of a PAL transmitter.

2.7 The SECAM system

The SECAM system is the third of the principal colour encoding systems in use today. It was developed in France in 1959 and was adopted there and in the USSR in 1966. The name SECAM is derived from the French *séquential couleur à mémoire* — that is, the use of sequential chroma signals and a memory device. The system has undergone some revisions since its conception. However, the following description explains the system as it is used today.

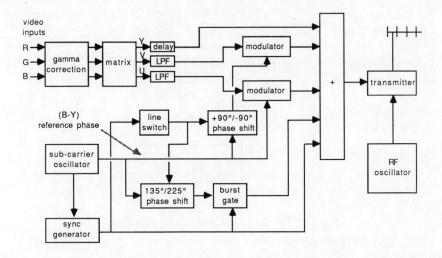

Figure 2.17   Block diagram of a PAL transmitter

The reasoning behind the development of this system is that if the horizontal resolution of colour transmissions can be reduced, then so also can the vertical resolution. If this is so, then it should not be necessary to transmit all the chrominance information on every scan line. Therefore why not transmit one colour difference signal on one line and the other on the next etc., thus avoiding the need for quadrature modulation and its inherent defects. However, in order to be able to reconstitute the original R, G and B signals, both chrominance signals must be available on every line. To achieve this, the chrominance information transmitted on one line is delayed and used again on the next scan line, resulting in a correct colour picture but with a reduced vertical colour resolution. The delay device used to achieve this in the receiver is similar to that used in the PAL system, but its accuracy does not have to be so great since a slight delay error only results in a colour registration error, not a hue error.

Having outlined the basis of the SECAM system it will now be described in more detail.

First, like the NTSC and PAL systems, the chrominance bandwidths are reduced to 1.5 MHz. Then, as mentioned above, the $(R' - Y')$ and $(B' - Y')$ colour difference signals are transmitted separately on alternate scan lines. Since only one chrominance signal needs to be transmitted during each line, quadrature modulation is not required, and other modulation methods can be considered. In order to obtain the inherent advantages of noise and interference performance, frequency modulation (FM) was chosen as the modulation method for the chrominance signals. Figure 2.18 illustrates the basic method of colour encoding in the SECAM system.

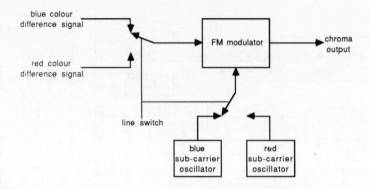

Figure 2.18 Basic SECAM colour encoding

The method of FM modulation involves some processing of the various signals which will now be explained.

2.7.1 Chrominance signal weighting

Like the NTSC and PAL systems, the gamma corrected $(R' - Y')$ and $(B' - Y')$ chrominance signals are weighted according to the proportions:

$$D'_R = -1.9(R' - Y') \qquad (2.44)$$

$$D'_B = 1.5(B' - Y') \qquad (2.45)$$

Notice the negative sign which implies that for a positive D'_R signal the FM carrier frequency is reduced and vice versa, whereas for a positive D'_B signal the reverse occurs.

2.7.2 Video pre-emphasis

As with most FM modulation schemes, pre-emphasis is applied to the signal before modulation. The form of pre-emphasis used in the SECAM system is defined by the expression:

$$G = 10\log_{10} \frac{1 + (f/f_1)^2}{1 + (f/3f_1)^2} \qquad (2.46)$$

where $f_1 = 85$ kHz and G is the relative gain as a function of frequency.

In rough terms, this gives a +4.6 dB amplitude increase at 100 kHz and +14 dB at 1 MHz. The result of this pre-emphasis is to introduce large overshoots at fast transitions of the chrominance signals which might cause problems in the modulator. To prevent this, the maximum amplitude of the chrominance signals is limited (or, more likely, the maximum deviations of the sub-carrier are limited instead).

2.7.3 FM modulation

After pre-emphasis, the chrominance signals are frequency modulated on to a colour sub-carrier whose undeviated frequency is near the top of the video band as in the NTSC and PAL systems. Linear modulation of the sub-carrier is employed — that is, the frequency deviation varies linearly with the signal amplitude and sign. In the original SECAM system, both signals used the same carrier frequency and deviations; however, because each chrominance signal has a different range of positive and negative values, the resulting asymmetrical sideband transmission of the FM signal led to a certain amount of distortion. Thus in the current system, both the undeviated sub-carrier frequencies and the deviation factors are different for each chrominance signal. The actual values used are shown in table 2.3.

Table 2.3 SECAM sub-carriers and deviations

Colour difference signal	Undeviated sub-carrier frequency MHz	Maximum positive deviation kHz	Maximum negative deviation kHz
Red	4.40625	350	500
Blue	4.2500	500	350

Figure 2.19 shows the carrier frequency ranges for the two colour difference signals — notice that a negative $(R' - Y')$ causes a positive sub-carrier frequency shift as indicated in equation (2.44).

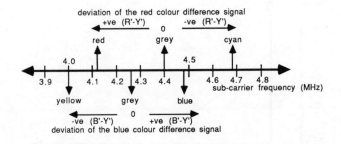

Figure 2.19 SECAM sub-carrier deviations

The sub-carrier is present at all times except during the field blanking interval and line sync pulses. It is however present during the identification lines in the field blanking interval (see later), and also during the back porch of each line sync pulse to allow the limiters in the receiver's FM demodulators to stabilise.

Finally, in order to minimise the subjective effect of the sub-carrier patterning on monochrome receivers, the phase of the sub-carrier is altered by 90° every third line and by 180° every field. This produces a pattern which repeats every three pictures, which is considered to be the least objectionable. It has no effect on the FM modulation, and no account need be taken of it in the receiver.

2.7.4 Sub-carrier pre-emphasis

In addition to the pre-emphasis of the chrominance signals, pre-emphasis is also applied to the modulated sub-carrier. It has the effect of increasing the amplitude of the carrier for increasing deviations. It is introduced to reduce the sub-carrier patterning on monochrome receivers while still retaining good noise immunity for saturated colours where it is most noticeable. Unlike the PAL and NTSC systems, the sub-carrier does not disappear during monochrome sections of a picture. Thus in order to reduce the interference effects, the sub-carrier amplitude is reduced as the saturation decreases (decreasing chrominance amplitudes). The exact form of this pre-emphasis is:

$$G = 10\log_{10} \frac{1 + 256(f/f_c - f_c/f)^2}{1 + 1.6(f/f_c - f_c/f)^2} \qquad (2.47)$$

where f_c is 4.286 MHz and G is the relative gain as a function of the deviation.

The actual shape of the pre-emphasis used is not critical since it only affects the interference effects, and not the resulting chrominance amplitudes (because of the limiters in the FM demodulator). Figure 2.20 shows the form of this pre-emphasis. Because of its shape, it is sometimes referred to as the *bell curve*.

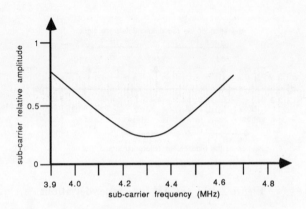

Figure 2.20   SECAM high-frequency pre-emphasis

Colour Television

The resulting FM modulated sub-carrier is then added to the luminance signal in the same way as in the NTSC and PAL systems. Luminance notch filters are also incorporated to prevent luminance information upsetting the FM chrominance demodulators in the receiver.

2.7.5 Chrominance identification signal

As with the PAL system, there must be an identification signal that will allow the receiver to detect which chrominance signal is being transmitted. Unlike the PAL system, however, this signal is not present on each line, but is included in every field blanking interval. Thus if the receiver chrominance switch is out of phase, then it will be one field period later before it can be corrected.

The actual identification signal is contained in nine line scans after the field equalising pulses. It consists of an FM modulated sawtooth waveform of the D'_B and D'_R sub-carriers on alternate lines. The sawtooth is negative on $(B' - Y')$ lines and positive on $(R' - Y')$ lines. It is produced by applying the required sawtooth waveform to the chrominance modulators, and thus results in an amplitude modulated as well as frequency modulated sub-carrier (because of the sub-carrier pre-emphasis). This signal therefore allows the receiver to check and adjust if necessary its chrominance switch phase during each field blanking interval. The form of the signal is illustrated in figure 2.21.

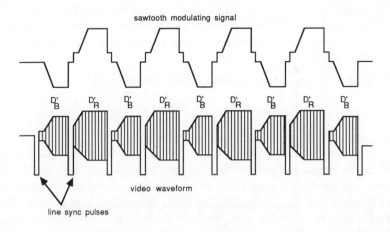

Figure 2.21 SECAM ident signal

2.7.6 Interference and compatibility effects

The monochrome compatibility is the same as for the NTSC and PAL systems, since the R, G and B signals are gamma corrected before the chrominance/luminance matrix, and the colour difference signals are also narrow band.

Patterning on a monochrome receiver from the sub-carrier is, however, slightly worse since, as mentioned above, even on monochrome pictures there must be some residual sub-carrier present. (Note that the broadcaster can deliberately switch off the sub-carrier for a fully monochrome programme; this can be detected by the receiver which will then switch off its colour decoding circuitry.)

Cross-colour however should be improved, as long as the amplitude of any luminance components in the chroma band are not large enough to upset the receiver's FM limiters.

2.7.7 Summary

(a) In the SECAM system, the two colour difference signals are transmitted alternately on consecutive scan lines. To reproduce the three colour signals on every line requires that each colour difference signal is used on adjacent scan lines. This results in a reduction of the vertical chrominance resolution.
(b) The colour difference signals are weighted and FM modulated on to a colour sub-carrier with pre-emphasis.
(c) The colour sub-carrier also has pre-emphasis applied, and different frequencies and deviation factors are used on the red and blue chroma signals. This is done to reduce monochrome patterning and to equalise as far as possible the frequency ranges of the FM chroma signal respectively.
(d) Identification of the chrominance signal on each line is achieved by inserting in nine lines after the field sync pulse an FM modulated sawtooth waveform, which is negative for blue colour difference lines and positive for red colour difference lines.

Figure 2.22 shows a block diagram of a SECAM transmitter.

2.8 Summary of the three colour encoding systems

This chapter has looked at how the composite monochrome video signal can be modified to produce a compatible and reverse compatible colour signal. The three principal methods of encoding the colour information have been described.

Under ideal conditions, a comparison of the three systems produces comparable results but an NTSC coded picture is noticeably inferior in practice, mainly because of the hue effects from phase errors. The other two systems perform equally well, but each produces different visual effects as a result of noise and interference, or even a misaligned decoder in the television receiver.

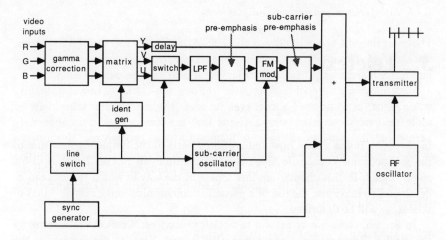

Figure 2.22 Block diagram of a SECAM transmitter

An added complication to the broadcaster of a SECAM signal is that of mixing pictures (fading between one picture and another etc.). This arises because of the chrominance FM modulation: the two signals cannot be simply mixed together as is the case with the chrominance amplitude modulation in the PAL and NTSC systems. Mixing in the SECAM system must be done by separating the luminance and chrominance signals before performing the mix or fade, and then recombining the result to produce the final SECAM signal.

It is important to note that at the outset of the development of the various systems one of the main requirements of a colour system had to be that of compatibility — both in terms of existing monochrome receivers and in terms of not increasing the existing video bandwidth. In order to achieve this, the methods adopted rely very heavily on the properties of the eye and the way in which we see images. Such factors as visual acuity, our ability to distinguish colour detail and the eye's varying sensitivity to different colours have all been exploited to the full in the development of each system.

If there were no compatibility constraints and the designers could start with a 'clean slate', then the systems would turn out quite differently.

3 Teletext

In chapter 2 it was shown how under-utilised parts of the frequency spectrum of the monochrome composite video signal are employed to create a compatible colour signal. In this chapter another aspect of the way in which systems can be developed to maximise the use of a particular communications channel — that of teletext — will be described.

In Britain, there are at present two teletext services: 'Ceefax' which is run by the British Broadcasting Corporation (BBC), and 'Oracle' the service of the Independent Broadcasting Authority (IBA).

Teletext is the name given to the system of transmitting information on previously unused parts of the composite video signal. Do not get confused between teletext and viewtex, which is an interactive system of transmitting information via telephone lines.

Teletext was designed in Britain in the early 1970s. At first, the BBC and IBA had different, incompatible, systems; but common sense prevailed and a single system was specified in 1974 — known as the 'White Book' specification. It was not until the early 1980s, however, when the British government introduced financial incentives, that the system began to be widely used and accepted — even though the broadcasters had been transmitting a teletext service for six or seven years.

Today (1988) the system has been adopted by many countries world-wide, and the specification has recently been extended to include many additional features and facilities. This chapter will, however, describe the specification contained in the original 'White Book' together with a few minor enhancements which were added in 1976. Chapter 6 will briefly describe the features, facilities and implementation of the new specification.

3.1 The basic teletext system

As mentioned above, teletext is a system whereby previously unused parts of the composite video waveform are used to transmit information. This information is decoded and the television picture tube in the receiver is used to display the resulting information, as illustrated in figure 3.1.

It is important to realise from the start that the final displayed teletext picture is generated by circuitry within the television itself in response to digital signals

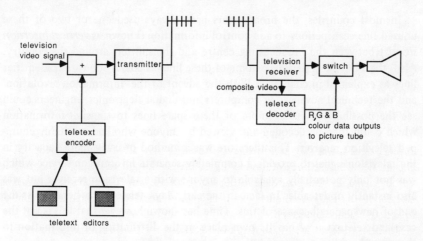

Figure 3.1 Basic teletext system

received. Thus in the presence of noise or other interference, the effect is to produce errors in the text of the received information, rather than a noisy or distorted picture — which would be the case with a normal television picture under these conditions.

It will be remembered that there are some unused line scan periods (that is, line scans during which there is no video information) after the field sync and equalising pulses and before the start of the next field scan which were originally included to allow the scanning circuits in the receiver time to get ready for the next field scan. This is illustrated in figure 3.2.

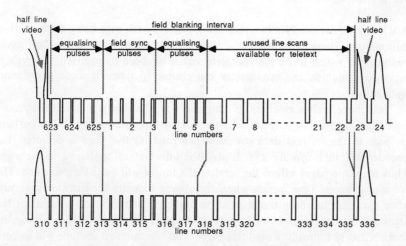

Figure 3.2 Unused line scans in 625-line system

In most countries, the broadcasters have always used one or two of these unused line scan periods to add control information (known as *vertical insertion signals*) between the broadcasting centre and transmitters around the country. Also, the SECAM system uses some of these lines as colour sub-carrier identifier lines as explained in chapter 2. With the advent of the 'information revolution' and the technical advances in computers and digital electronics, engineers could see the possibility of using more of these spare lines to transmit information which could then be decoded and viewed by anyone who had a suitably equipped television receiver. This therefore was a method of using spare capacity in the television signal to provide a completely separate information service which was not only potentially available to anyone with a television receiver but was also instantly updateable. In fact, in the early days, teletext was heralded as the end of newspapers because of this. Time has shown however that this is not the case and teletext now has its own place in the distribution of information to the general public.

In essence, therefore, the teletext system is a method of using some of the spare line scan periods before the start of each picture scan proper to include information which can be decoded and displayed on the television screen. This information is potentially available to anyone who can receive the associated television signal and has a suitably equipped television set or separate decoder.

In order to study the detailed operation of the teletext system, it is helpful to adopt a top-down approach. Thus the organisation of a teletext service and the display characteristics will first be described before the technical details of the teletext signal.

3.2 Service organisation

A teletext service consists of a number of *pages* — each page being a screenful of information. These pages are transmitted serially along with the television composite video signal. When the complete service has been transmitted, the cycle is repeated (although the broadcaster can choose to transmit some pages more frequently).

The service, unlike videotex, is thus one-way; the user is unable to request a page directly, but can only instruct the teletext decoder to search for a particular page in the teletext data stream. When (and if) this page is detected, the decoder will then capture and display the information contained in the page. Thus the more pages within the service, the longer will be the access time. The worst-case access time occurs when the viewer requests a particular page just after that page has been transmitted on one cycle and thus has to wait for the transmission on the next cycle before the picture is displayed (assuming the receiver has not already stored this page). The minimum access time will be zero since the page request might occur just as the required page is starting to be transmitted. Thus the average access time will be half the cycle time.

3.2.1 *Page addressing*

Clearly there must be some addressing contained within the teletext data to indicate to the decoder in the receiver what page and what row of that page is being received (and in a general case what column as well, although the serial nature of the service means that rows are filled sequentially from left to right — thus there is no need for column addressing).

A teletext service is divided into up to eight *magazines*. Each magazine can contain up to 100 pages. Pages from different magazines can be interleaved, but different pages from the same magazine cannot. The reason for this will become clear later.

In addition to the magazine and page addressing data, each page has an associated *sub-code*. This code was originally intended to carry time information so that pages could be time coded. However, this feature is rarely used, and the extra sub-code addresses can be used to extend the addressing range to many thousands of individual pages within each magazine. The detailed range of these sub-codes will be discussed later.

3.3 The teletext display

What does a teletext page look like on the television screen, and how is it formed?

A teletext display consists of text (letters, numbers etc.) and simple graphic shapes. In addition, there are a number of control codes which allow selection of graphic/text colours and other display features known as *attributes*. Each character or graphic shape occupies a defined area of the screen called a *character rectangle*.

3.3.1 *Determination of character size*

What factors determine the number of characters that can be fitted across and down one teletext page (that is, how many *columns* and *rows*)?

At the time when the teletext system was being designed, computer displays usually used a minimum of 80 columns per row. However the resolution of a domestic television set is insufficient to allow this many characters to be satisfactorily displayed along one row. In addition, a person looking at a teletext display from an average television picture viewing distance would find it difficult to read the 80 column text.

Taking into account these considerations, it was decided to use 40 characters per row. From similar considerations, the number of rows was determined, and a value of 24 was specified — although as we shall see later, an extra 25th row is used in the new enhancements to the specification.

3.3.2 Display characters

The characters available for display in the teletext system are the letters A to Z and a to z, numerals 0 to 9, and the common punctuation marks and symbols (currency signs etc.). The new specification, however, now allows accents and a whole range of other character sets and symbols to be displayed.

As mentioned above, the system also allows for a certain number of graphic shapes to be displayed which enable simple pictures to be created. The resolution of these shapes is however not great, and only rough pictures with rectangular outlines can be created. Originally the system was designed to be used for text information, and as such pictures were not considered important — the graphic shapes being included only to allow for the display of large headings and simple diagrams. They are thus usually termed *block graphics*.

To obtain these block graphic shapes, a character rectangle is divided into a matrix of 2 x 3 smaller rectangles. Each of these small rectangles can either be illuminated or blanked, and there is a direct correspondence between the display of these small rectangles and the state of individual bits in the transmitted code. This is illustrated in figure 3.3.

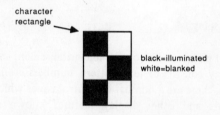

Figure 3.3   Typical teletext block graphic

Since the system was designed when colour television was in existence (albeit in its early days) coloured teletext displays were catered for from the outset. These and various other display attributes, such as flashing and double height, are invoked by a number of control codes.

There are seven colours available — the basic primary colours red, green and blue, together with the three combinations of two of these colours (yellow, magenta and cyan) and white, produced by combining all three primary colours.

Every control code occupies a character rectangle but is displayed as a space. The effect of the code persists until the end of the current row unless it is cancelled by the inclusion of a further suitable control code. This means that whenever there is a need to change the colour or other attribute of the display, a space must occur for the required control code. Thus it is not possible, for example, to display adjacent letters in different colours. At first sight this appears to be a great disadvantage, but again, since teletext was designed as an information service, this was not considered too much of a limitation. The new specification does however include methods to produce such displays. Because of the position of these control codes within the display (and the data stream), they are usually termed *serial* or *spacing* attributes.

Figures 3.4 and 3.5 are sample teletext displays (in black and white only) showing the graphics and text capabilities of the system.

Figure 3.4 Sample teletext display (in black and white)

3.3.3 Character coding

Each character and control code is represented for transmission in the video signal as a unique 7-bit code. The codes used are based on the ASCII (American Standard Code for Information Interchange) code which is almost universally used for computer communications.

The ASCII code table has 128 different codes and thus a 7-bit binary number is required to specify any individual code. The table is usually drawn as a grid of 16 rows and 8 columns. A particular row is defined by the lower 4 bits of the code, and the column by the upper 3 bits. Figure 3.6 shows the full teletext code table.

Figure 3.5   Sample teletext display (in black and white)

Columns 0 and 1, which are used for control codes in computer communications, contain the teletext serial attribute controls. Notice that some positions are undefined. Columns 2 and 3 contain two variations; at the start of each row or after an alphanumeric control code, the numerals or punctuation marks from column 2a/3a are displayed. However, if there has been a previous graphics control code, then the graphics characters from column 2b/3b are used. Columns 4 and 5 contain the capital letters and a few other symbols — these are always displayed as these letters/symbols, even if there has been a previous graphics control code (displays containing capital letters in the middle of a graphic display are sometimes called *blast-through* alphanumerics). Finally, columns 6 and 7 contain two variations — the small letters and a few more symbols in alphanumeric mode (columns 6 and 7) and the remaining graphics symbols in graphics mode (columns 6a and 7a).

Quite often, individual characters are referred to by the shorthand 'X/Y' where X is the relevant column and Y the row in that column at which the character is to be found. For example, capital A has a code of 4/1, which corresponds to a binary code of 1000001_2 or a hexadecimal code of 41_{16}. This convention will be used in this book, as will the subscript convention for indicating the base of numbers (for example, 2, 8, 10, 16 for binary, octal, decimal, hexadecimal respectively) as above.

Teletext

column	bit numbers 6 5 4	0 0 0	0 0 1	0 1 0	0 1 1	1 0 0	1 0 1	1 1 0	1 1 1
row 3 2 1 0		0	1	2 / 2a	3 / 3a	4	5	6 / 6a	7 / 7a
0 0 0 0	0			space	0	@	P	_	p
0 0 0 1	1	alpha red	graphics red	!	1	A	Q	a	q
0 0 1 0	2	alpha green	graphics green	"	2	B	R	b	r
0 0 1 1	3	alpha yellow	graphics yellow	£	3	C	S	c	s
0 1 0 0	4	alpha blue	graphics blue	$	4	D	T	d	t
0 1 0 1	5	alpha magenta	graphics magenta	%	5	E	U	e	u
0 1 1 0	6	alpha cyan	graphics cyan	&	6	F	V	f	v
0 1 1 1	7	alpha white	graphics white	'	7	G	W	g	w
1 0 0 0	8	flash	conceal	(	8	H	X	h	x
1 0 0 1	9	steady	contiguous graphics	)	9	I	Y	i	y
1 0 1 0	10	end box	separated graphics	*	:	J	Z	j	z
1 0 1 1	11	start box		+	;	K	←	k	¼
1 1 0 0	12	normal height	black background	,	<	L	½	l	‖
1 1 0 1	13	double height	new background	-	=	M	→	m	¾
1 0 1 0	14		hold graphics	.	>	N	↑	n	÷
1 0 1 1	15		release graphics	/	?	O	#	o	■

Figure 3.6 Teletext code table

As mentioned before, all of the control codes when received will be displayed as a space character (2/0). Most of the codes are self-explanatory, but some need a little explanation.

FLASH (0/8) causes subsequent characters/graphics to flash on and off at approximately $\frac{1}{2}$ second intervals. STEADY (0/9) cancels this effect.

START BOX (0/11) causes the start of a 'box' of text within the display of the television programme. END BOX (0/10) finishes this box.

CONCEAL (1/8) causes subsequent text to be hidden (displayed as spaces). The text is revealed by a user control. This effect is cancelled by any alphanumeric or graphic colour control.

SEPARATED GRAPHICS (1/10) causes the graphics characters to be displayed so that each block in the 2 x 3 matrix is displayed smaller and therefore becomes separated from the next block. This effect is cancelled by the CONTIGUOUS GRAPHICS (1/9) code. These codes have no effect in alphanumeric mode.

NEW BACKGROUND (1/13) causes the background colour of the remainder of the row to take on the colour of the current character/graphic display.

HOLD GRAPHICS (1/14) causes the graphics character from the previous character rectangle to be displayed instead of the space character for a control code. This allows graphics pictures to be built up with abrupt colour changes (that is, the colour change control code is 'hidden'). RELEASE GRAPHICS (1/15) cancels this effect.

Finally, it was mentioned above that the display of each graphic block in the 2 x 3 character rectangle matrix is controlled by one bit in coded data. Figure 3.7 shows the bit numbers of the 7-bit binary code that controls each block in the character rectangle. It can easily be confirmed that any particular combination of bits does in fact give the corresponding graphic display as shown on figure 3.6. Notice that bit 5 is always set to 1 for a graphic character.

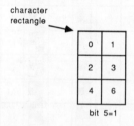

Figure 3.7   Graphic character/bit correspondence

3.4 The teletext TV line

To keep the decoding simple, the data for one teletext display row is contained together with its addressing data in the active part of one line scan interval. This combination of data is usually called a teletext *packet*. In the 625-line system there are 17 available line scans after the field equalising pulses and before the first picture line scan, although some are already used for the vertical insertion signals and (for SECAM) the colour identification signal.

3.4.1 Data encoding

The data for each teletext row is transmitted serially as 8-bit bytes. The main display data uses seven of these bits for the display code as explained previously, and the eighth bit is used as an odd parity bit. In the event of a parity error being detected, the character is not written to the screen. The addressing data uses a more sophisticated form of error protection which will be explained later.

Various methods of forming the serial data stream were considered; some included clock information to simplify receiver design but required more bandwidth, and others used less bandwidth but did not include clock information. The method chosen involves a 2-level binary signal, the high level corresponding to a logic 1 and the low level corresponding to logic 0. Each signal bit occupies equal time intervals within the teletext packet. This method is called *non-return to zero* (NRZ) coding, and it is illustrated in figure 3.8 which shows the serial data stream for the letter 'S' — note the setting of the parity bit to achieve an odd number of bits in the complete 8-bit word. In all cases, the least significant bit is transmitted first in time.

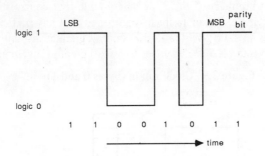

Figure 3.8 Serial data for the letter 'S'

3.4.2 Decoder synchronisation

The disadvantage of this method of coding is that it does not include any clock information, neither are there any start or stop bits to indicate where each 8-bit byte starts (as there commonly are in asynchronous data transmissions via telephone lines). Thus the receiver must have some way of (a) detecting the start of each bit (there will not be a level transition before each bit if the code contains two or more of the same logic level together) and (b) the start of each byte — that is, where in the bit stream one byte ends and the next commences. These two actions, known as *bit synchronisation* and *byte synchronisation* respectively, are achieved by prefixing the serial data stream with extra data bytes as follows.

The first two bytes in the packet contain alternate ones and zeros as shown in figure 3.9 — this provides the basic bit rate of the data. A local oscillator is phase-locked to this signal, and is arranged to remain sufficiently in phase to enable the remainder of the teletext packet to be sampled at the correct time in order to recover the data successfully. This provides the bit synchronisation information for the receiver. It is the digital equivalent of the colour sub-carrier burst used to synchronise the local sub-carrier oscillator in the television's colour decoder.

After these first two bytes comes a data byte containing a unique code which allows the decoder to detect the start of each individual byte of data. The code specified is 27_{16} and is known as the *framing code* as illustrated in figure 3.10. The decoder, having achieved bit synchronisation, starts to sample the data stream until it detects the framing code byte, at which point byte synchronisation has been achieved and decoding of the main data stream can commence.

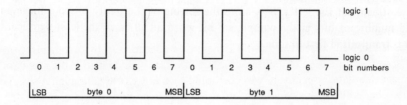

Figure 3.9 Clock run-in (bytes 0 and 1)

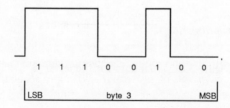

Figure 3.10 Framing code (byte 2)

3.4.3 Addressing data

After the three synchronising bytes, the data proper starts with the addressing information. From the previous explanation of the organisation of a teletext service, it can be deduced that for each packet the decoder must have available magazine, page, sub-code and row addressing data. In addition there is a set of control bits which are used to indicate to the receiver various details about the form of the teletext page being transmitted. These are details such as whether the page is a normal text page or a subtitle/newsflash page where the data is to be boxed on to the television picture.

However, to save on the quantity of data transmitted, general display packets only contain magazine and row addressing data. The remainder of the addressing data is contained in the data for the top row (row 0) of every page. This row has a special format and is called the *header row*. These header rows are used to 'introduce' a page, and any subsequent packets with the same magazine number are deemed to belong to that page until the next header row is encountered. Thus pages in different magazines can be interleaved, but pages within the same magazine cannot. We will look at the form of these special header rows later.

For teletext display rows other than row 0 (that is, the data contained in packets with row addresses of 1 to 23 – usually called *packets 1-23*), the next two data bytes (bytes 3 and 4) after the three synchronising bytes contain the magazine and row address data. It is clearly important that the data in these bytes is received correctly to avoid either rows appearing in the wrong place on the screen or in a page of a different magazine. This would occur if the row data or magazine data respectively was corrupted by a noisy or weak signal. A simple parity check bit on this data therefore is not sufficient, and a more sophisticated method of error protection is employed. The method used involves an error protection bit for each data bit, and these bits are coded using a *Hamming error-correcting code*. This coding scheme allows single bit errors in each byte to be corrected, and even numbers of errors (except 8) to be detected but not corrected, but at the expense of doubling the number of bits that have to be transmitted.

3.4.4 Hamming error protection

The subject of error protection is complicated and there are many schemes available to achieve different levels of protection. With any scheme there is a trade-off between the level of protection (and correction) of the data that can be achieved and the additional protection data that has to be transmitted. The Hamming-coded data used in teletext transmissions is formed as follows.

Numbering the eight bits in each byte 0 to 7, the protection bits in the Hamming coded byte are interleaved such that bits 1, 3, 5 and 7 are the message bits (bit 1 = least significant data bit) and the protection bits are the even numbered bits.

The protection bits are set to the result of parity checks on different sets of bits as below:

> bit 0 is set for odd parity over bits 0, 1, 5 and 7
> bit 2 is set for odd parity over bits 1, 2, 3 and 7
> bit 4 is set for odd parity over bits 1, 3, 4 and 5
> bit 6 is set for odd parity over all bits

On reception, four odd parity checks are carried out on the data as below:

> check A on bits 0, 1, 5 and 7
> check B on bits 1, 2, 3 and 7

check C on bits 1, 3, 4 and 5
check D on all bits

The result of check D enables all even numbers of errors to be immediately detected, and the result of the other checks enables single bit errors to be detected and corrected.

Table 3.1 contains the action that the decoder should take for each of the sixteen possible combinations of the results of the four parity checks.

Table 3.1 Hamming parity check decoder actions

Parity checks				Decoder action
A	B	C	D	
0	0	0	0	OK - no errors
1	0	0	0	)
0	1	0	0	)
1	1	0	0	)
0	0	1	0	) reject - even no. of errors
1	0	1	0	)
0	1	1	0	)
1	1	1	0	)
0	0	0	1	complement bit 6
1	0	0	1	complement bit 0
0	1	0	1	complement bit 2
1	1	0	1	complement bit 7
0	0	1	1	complement bit 4
1	0	1	1	complement bit 5
0	1	1	1	complement bit 3
1	1	1	1	complement bit 1

0=parity check OK
1=parity check failed

3.4.5 Magazine and row address group

As mentioned above, bytes 3 and 4 in every teletext packet contain the magazine and row address, Hamming encoded. Since up to 8 magazines are allowed, three data bits are required. Also, in order to specify one of 24 rows, a 5-bit binary code is required. These two addresses are combined with the various Hamming protection bits to form bytes 3 and 4 as shown in figure 3.11.

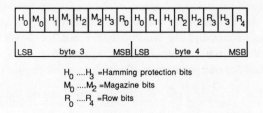

H_0 ....H_3 = Hamming protection bits
M_0 ....M_2 = Magazine bits
R_0 ....R_4 = Row bits

Figure 3.11 Magazine and row address bytes

3.4.6 Teletext display packets 1-23

Following the magazine and row address group in packets 1-23 is the display data proper. This consists of 40 bytes corresponding to the 7-bit + odd parity codes that represent the information to be displayed on the 40 columns of the specified row from the left to the right of the screen. After parity checking, this data, if required, is stored directly in the decoder's display memory ready to be interpreted and displayed.

Figure 3.12 shows the format of a complete teletext packet for packets 1-23.

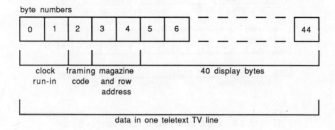

Figure 3.12   Format of packets 1-23

Thus, after a teletext page has been introduced by a header row, subsequent television scan lines available for teletext data contain the packets required to specify the rest of the page completely. Alternatively, it is possible to interleave packets from different magazines on consecutive television scan lines. In both cases, the completion of the transmission of a page is indicated by the transmission of another header row with the same magazine number. This means that it is only necessary to transmit packets for rows that contain some display or rows that need updating – packets for blank rows need not be transmitted. Also, packets need not be transmitted in row sequence, although they usually are.

3.4.7 Header row format

Finally, we have seen that a page is introduced by a header row which has a different format to the other rows. This header row is transmitted as packet 0 and contains the page and sub-code addresses, control data and the display for the top row of the page (row 0). The format of this row is as follows.

Bytes 0 to 4 contain the clock run-in, framing code and magazine and row address group like all other packets. Byte 5 contains the page number units in BCD format and Hamming protected. Byte 6 contains the page number tens in similar format. The next four bytes contain the page sub-code address, again in BCD format and Hamming protected. Because these addresses were originally intended for time-coding purposes, each sub-code digit has a different number of bits allocated. The spare bit(s) in any group of four data bits are used as control bits. Figure 3.13 shows the format of the page and sub-code address byte.

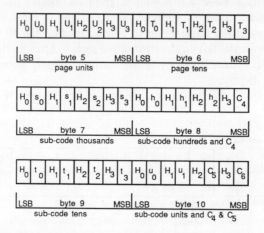

Figure 3.13 Page and sub-code address bytes

There then follow two further Hamming-protected bytes that contain eight more control bits. In the original specification some of these bits are reserved for future use.

Finally, the remainder of the packet (bytes 13 to 44) contain the display data for row 0 starting at column 8. Although this data can contain any display, it usually consists of the service name, page number as ASCII characters, the current day, date and time. When a user selects a page, all headers received are usually displayed directly so that the user sees the numbers of the pages being transmitted. This display is called a *rolling header display*.

To summarise, figure 3.14 shows the complete format of packet 0.

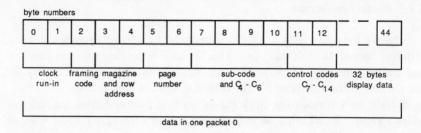

Figure 3.14 Format of packet 0

3.4.8 Header control bits

What are the function of the control bits? In the original system they were allocated as follows:

Control bits C_1, C_2 and C_3 are the three magazine number bits and are now usually referred to as the *magazine*. The remaining control bits, when set, have the following functions.

C_4 indicates that the display should be cleared because the page data has been changed. If this bit is set, at least one frame period will elapse before any further packets of this page are transmitted − this allows the decoder time to clear all the data out of its display memory.

C_5 indicates that the page is a newsflash page − that is, the television picture should be displayed and text is to appear boxed within it (when indicated by the required box on/off control codes).

C_6 indicates a sub-title page; its effect is the same as the newsflash bit above.

C_7 instructs the decoder to suppress the rolling header display for this page. This is done typically if a page is transmitted more frequently than others, and is thus not in sequence with the others.

C_8 indicates that the page has been updated in some way − this might cause, for example, the newsflash display to be immediately shown boxed into the television programme.

C_9 indicates that the page is not in sequence.

C_{10} instructs the decoder not to display the page directly. This is used, for example, for telesoftware pages (pages containing computer programs etc.) which are unintelligible to humans.

C_{11} indicates that magazines are not interleaved − magazines are sent serially.

C_{12}-C_{14} are reserved for future use. In fact they are used in the new specification to invoke different character sets.

Every page contains the above set of control bits in its header row, and thus different types of pages (normal display, newsflash, subtitle etc.) can be interleaved at will. Notice the inclusion of the one frame period delay after the header of a page with C_4 set to enable time for page clearing.

3.5 The teletext signal

Having looked at the logical structure of the teletext TV line, it can now be shown in physical terms how this is fitted into the unused line scans in the field blanking interval.

The teletext signal commences immediately after the back porch of the line sync pulse corresponding to the start of a line period available for teletext. The logic 0 level is represented by the video black level and the logic 1 state is represented by a level of 66 per cent of the difference between black level and peak white level as illustrated in figure 3.15.

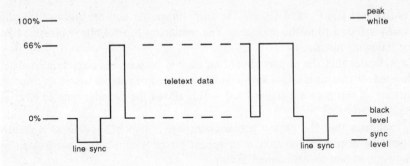

Figure 3.15 Teletext data signal levels

Figure 3.16 is an oscilloscope trace of a typical teletext television scan line and figure 3.17 is an expanded trace of the start of such a scan line. This figure shows the end of the sync pulse, the back porch (without burst), the clock run-in and framing code and the start of the addressing/display data.

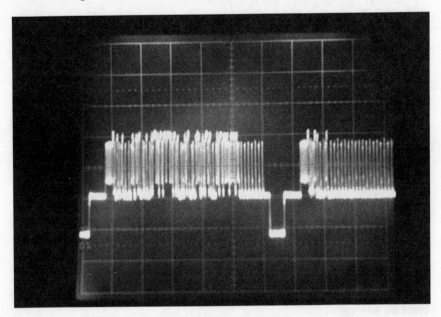

Figure 3.16 Typical teletext television scan line

Teletext

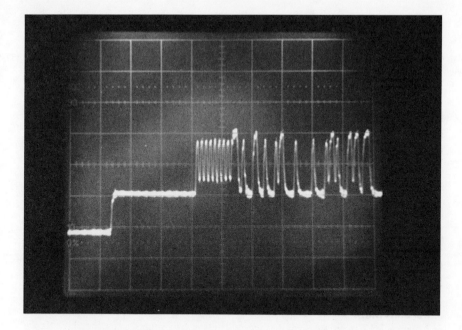

Figure 3.17 Start section of a teletext television scan line

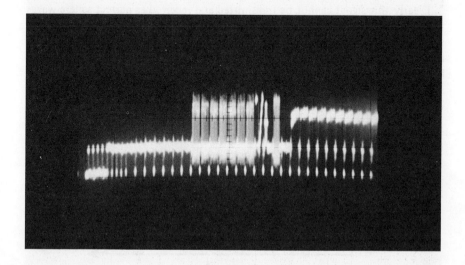

Figure 3.18 Teletext data in the field blanking interval

Television and Teletext

Figure 3.18 is an oscilloscope trace of the vertical blanking interval. It shows the field sync pulses, the teletext data scan lines, the vertical insertion signals and the start of the visible picture scan.

3.5.1 Bit rate

As already noted, the teletext data line consists of forty-five 8-bit bytes (including the clock run-in, framing code and magazine/row address group). This data must be fitted in the active line period between line sync pulses.

In a 625-line system, the active line period is 52 μs. Thus the minimum bit rate required to accommodate the 45 data bytes in this time is given by

$$\frac{45 \times 8}{52 \times 10^6} = 6.923 \text{ Megabits/s} \tag{3.1}$$

To allow for a small margin of error, the actual bit rate chosen is 6.9375 Mbits/s which is 444 × line frequency. From similar considerations, the bit rate for 525-line systems is 5.727 272 Mbits/s (or 364 × line frequency).

3.5.2 Bandwidth

The frequency components of the digital signal shown in figure 3.15 theoretically extend to infinity for a signal containing alternate ones and zeros. In practice, significant harmonics extend up to some 40 MHz with a fundamental frequency of half the bit rate — that is, 3.468 75 MHz for the 625-line system.

Clearly this signal cannot be transmitted and received as perfect square waves since the bandwidth of the 625-line television system is only 5.5 MHz and less in the 525-line system. To prevent these high harmonics from affecting the RF modulators, the logic signal is passed through a low pass filter which reduces the harmonic components to essentially zero by 5 MHz. The resulting frequency spectrum of the teletext data is shown in figure 3.19, and its approximate form is illustrated in figure 3.20. This waveshape is usually known as a *raised cosine*.

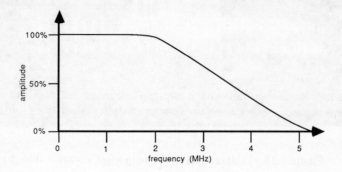

Figure 3.19 Teletext data frequency spectrum after filtering (625-line system)

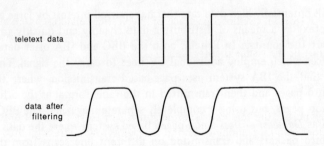

Figure 3.20 Teletext raised cosine data waveform

3.5.3 Data rate

Since the teletext data is only transmitted in a few line periods during the field blanking interval, the average data rate will be much less than the bit rate described above. Also, each teletext data line consists of the 'overheads' of clock run-in, framing code and address data, which will reduce the average data rate further.

The average data rate can however be deduced by considering the number of lines per frame in which teletext data is carried. This calculation will assume for simplicity that the extra addressing and control information in the header row is not counted as 'overhead' data. Thus in each teletext data line there are 45 − 5 = 40 bytes of actual display data.

If the field period of the system is T_f, and in each field N lines are used to carry teletext data, then the average data rate will be given by

$$\frac{40 \times 8 \times N}{T_f} \text{ bits/s} \tag{3.2}$$

For the 625-line system where T_f is 20 ms, then for a broadcaster using, say, four lines per field for teletext data, the average data rate will be

$$\frac{40 \times 8 \times 4}{20 \times 10^{-3}} = 64 \text{ kbits/s} \tag{3.3}$$

Compare this average rate with a common telephone line data rate of 1200 bits/s, and it can be seen potentially how much faster the data rate of teletext can be.

At the time of writing (1988) the British broadcasters are using up to eight scan lines for their main teletext service — remember however that if a broadcaster has two or more television channels, each one can have a different teletext service associated with it.

It is this potentially high data rate that has spurred interest by large companies using teletext as a means of distributing data cheaply and quickly to branches throughout the country. In Britain, both the BBC and IBA offer data distribution services which employ additional TV lines to carry the signal. The systems differ in that the IBA system uses page-based transmission, where the data is formed into pages and then transmitted in the same format as the other normal information pages, but using a completely separate magazine. The BBC's system – known as *Datacast* – uses a non-paged based system where the data is formed directly into packets and transmitted on different line scans from the Ceefax data.

3.6 Full channel teletext

One of the main criticisms of the teletext system has always been the long access time. The broadcasters usually adjust the size of their services and the number of scan lines used to obtain a cycle time of around 30 seconds – giving an average access time of 15 seconds. This may not sound very long, but it can be very frustrating for users who need to look at a large number of pages.

One of the ways of reducing the apparent access time is by incorporating more page stores (memories) in the decoder, so that the decoder can store frequently needed pages such as indexes ready for instant access. This is the basis of the *Fastext* enhancement which will be discussed in chapter 6.

Another exciting possibility which is being considered is the use of *full channel teletext* where all line scans (apart from the field sync and equalising pulses) are used to carry teletext data – that is, there is no television picture and all the active scan lines as well as those in the field blanking interval contain teletext data.

To get a rough idea of the theoretical data rate of a full channel 625-line system, assume that there are 300 lines in every field (600 per picture) that can contain teletext data; then, from equation (3.2) the average data rate is given by

$$\frac{40 \times 8 \times 300}{20 \times 10^{-3}} = 4.8 \text{ Mbits/s} \tag{3.4}$$

If there are a full 24 rows in each teletext page, then in each frame period 300/24 = 12.5 pages will be transmitted. Thus the page rate will be 12.5 × 50 = 625 pages per second! This is clearly an enormous data rate, and although it means dedicating a television channel solely to teletext, it is likely that such systems will be appearing in the not too distant future. The most likely use of full channel teletext will be on satellite broadcasting where there is plenty of bandwidth available.

It must be remembered though that the system is still one way, and can only be used for the general distribution of data, not as an interactive communications system.

4 Videotex

This chapter will look briefly at the videotex system since it has much in common with the teletext system. *Videotex* is the name given to any computer system that uses television screens to display information transmitted via telephone lines. A videotex system also has a certain method of structuring the information on the computer and a teletext-like method of displaying the data. The name videotex has now superseded the original name of *Viewdata* which was used by the then British Post Office (now British Telecommunications (BT)) for the first of these types of information systems.

Videotex was developed by the British Post Office in the late 1970s and the first edition of the specification was issued in January 1981. Their information service was (and still is) called *Prestel* and was the first such system to be operational. The original reason for the development of such a system was to increase the use of the telephone network by providing a consumer-oriented information service whose display was compatible with teletext displays. A videotex decoder could thus be easily and cheaply made by using the display circuitry from a teletext decoder.

Another important event in the development of videotex was in the *modem* − the device that converts the signals transmitted down the telephone line into digital signals, and vice versa. Until the first viewdata decoders were built, modems were constructed in expensive separate boxes and had to be supplied by the telecommunications authority. Videotex modems, however, although they still had to be approved by the authority, were built into the decoder itself (or into the television cabinet for an integral decoder) − this allowed the development of much cheaper modems and thus brought the cost of a videotex decoder down to a reasonable level for the consumer market.

Since the early 1980s, many other countries have adopted the original viewdata specification, or have developed their own similar system. The specification itself, like teletext, has also been upgraded and has now been adopted as an international standard.

This chapter will explain the features of the original specification − most of which are common to the various systems around the world. Chapter 6 will look briefly at the new features and facilities offered by the now international CEPT upgraded specification.

4.1 The basic videotex system

A videotex system is basically a time-share computer system which is specifically designed to hold a large volume of data, and can cope with a large number of simultaneous users. For a large system, the 'computer' may consist of a network of computers around the country which are connected by high-speed data lines.

Users access the computer via the normal telephone lines using full duplex serial data transmission and there is thus a two way communication between the user and computer. The data is displayed on a normal television screen using the same display format as the teletext system. The videotex decoder can either be integral in the television receiver cabinet, or in a separate box.

Data is input into the computer from 'information providers' using a modified videotex decoder, or by a special fast 'bulk update' method from the information provider's local computer.

Figure 4.1 shows a simple block diagram of the basic videotex system.

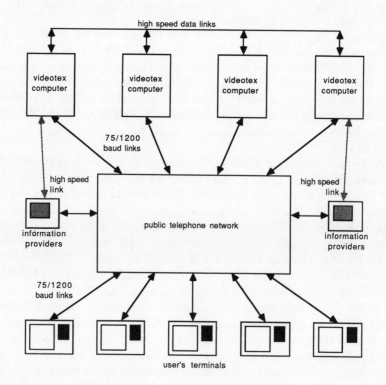

Figure 4.1 Basic videotex system

4.2 The videotex database

A videotex database, like teletext, is divided into a number of pages of data — each page being a screenful of information. Pages are also numbered, but a page can be assigned a number composed of any reasonable number of digits. In addition, any page number can have up to 26 associated sub-pages designated by the letters a to z.

Since videotex systems were designed to be compatible with teletext, selection of any page had to be made possible by the use of numeric keys only (any teletext page can be accessed by keying a three digit number).

Videotex systems do however employ two additional keys, labelled * and # (known variously as 'square', 'hash' or 'number').

4.2.1 Tree structure

Videotex pages are linked by a 'tree structure' whereby each page has a number of links to related pages. In a simple system, up to 11 links are provided; these are selected by keying one number (0-9) or the # key. The numeric keys can select any page which the database editor specifies, but the # key always selects the next sub-page. For example, if page 100a is displayed, keying # will cause page 100b to be displayed.

In a videotex database, therefore, the initial page (usually page 0a) will contain an index to the main sections of the database, and this allows the user to select further indexes and eventually the required information can be obtained. This process is illustrated in figure 4.2.

If the page number of the information is known, then this process can become very laborious, so by keying *page number# the required page (with sub-code a) can be selected directly. Also, the computer can remember a number (usually 4) of previous selections made by a user and each previous page can be retransmitted and displayed by keying *#.

Finally, in case the information is corrupted by interference or noise on the telephone connection, two methods of causing the page to be retransmitted are provided. Keying *09 causes the page to be retransmitted without any updates or extra charge, while keying *00 causes the page to be transmitted with any updates, but the page will be charged for again.

4.2.2 Charging structure

Unlike teletext, the system is interactive in that the user can instruct the computer to transmit a particular page of information. This means that, if required, the user can be charged for the information received.

In fact there are three components to the charge for accessing a videotex database. The first is the normal telephone charge (assuming normal telephone

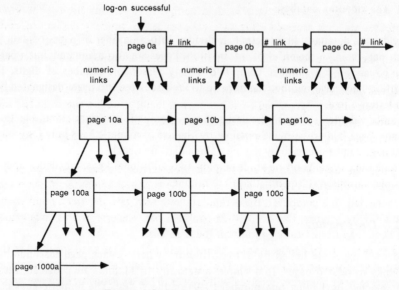

Figure 4.2 Videotex tree structure

lines are used); the second is a time charge for using the computer; and the third is a charge for the actual information received.

Although the vast majority of pages are usually free, the other two charges, especially if the computer is accessed at peak times, have contributed to the relatively slow growth in the number of videotex decoders in the consumer sector. It is in the business area that videotex systems have seen most growth — especially in the financial and related areas.

4.3 The videotex display

Videotex was specifically developed to be as far as possible compatible with teletext both in the codes used for the data and in the display itself. Hence, the videotex display contains 24 rows of 40 columns. The character set and block graphics are the same and the same set of serial attributes is used. However, there are some extra control codes which allow a cursor to be moved about the screen and the decoder to be remotely programmed with a user number, password and the telephone numbers of videotex computers to which the user is registered.

4.3.1 Videotex control codes

In the teletext system, columns 0 and 1 of the code table are used to transmit the serial attribute controls. In videotex, however, these columns are retained for the standard control functions used in computer communications. The serial attributes are thus accessed using an 'escape sequence' whereby the control code ESC (1/11) is sent first followed by a code from columns 4 or 5 of the code table. These codes access the serial attributes from columns 0 and 1 respectively of the teletext code table.

If the code after the ESC code is from column 3 rows 1-4, then the decoder can be remotely programmed (if it has such a facility) with a user number and password and/or one or more telephone numbers. These details are stored in a special non-volatile memory so that the information is retained when the decoder is switched off.

When the user wishes to access a videotex service, he can then instruct the decoder automatically to dial and log into the required computer. When a user is connected to a computer, the computer can interrogate the user's number and password by sending the ENQ (0/5) control code. Receipt of this code causes the decoder automatically to transmit these details.

The other control codes from columns 0 and 1 that have been specified in the original system concern the cursor. The cursor indicates the position on the screen where the next character received will appear. This position is called the *active position*, and the cursor usually appears as a flashing underline or reverse video character (that is, the background and foreground colours are interchanged) in this character rectangle. The cursor can be switched on or off with the control codes 1/1 and 1/4 respectively and moved up, down, left or right by one row/column by codes 0/11, 0/10, 0/8 and 0/9 respectively. It can also be 'homed' (returned to row 0 column 0) by code 1/14 and 'returned' to the start of the current row by code 0/13.

Finally, code 0/12 causes the screen to be cleared. Like teletext, a delay is usually incorporated after transmission of this code to allow the decoder time to perform the clear screen function — this delay can be obtained by transmitting a number of NULs (code 0/0) which are ignored by the decoder.

Figure 4.3 shows the complete videotex code table from the original specification. Note the blank control codes in columns 0 and 1 which are retained for compatibility with other computer communication systems.

4.4 The videotex signal

The videotex character and control codes are converted to serial format for transmission along the telephone lines. The serial data is coded into asynchronous start/stop bit format, and a simple even parity bit is used for error protection. Each serial data word therefore consists of 10 bits — a start bit, 7 data bits, a parity bit and one stop bit. The start bit is a logic 0 and the stop bit is a logic 1. Between each set of 10 bits comprising a word there can be an arbitrary delay of logic 1 before the start of the next 10 bit sequence. Figure 4.4 shows the serial data stream for the letter 'S'.

Bits	b7 b6 b5	0 0 0	0 0 1	0 1 0		0 1 1			1 0 0		1 0 1		1 1 0		1 1 1	
b4 b3 b2 b1	Row \ Col	0	1	2	2a	3	3a	3b	4	4b	5	5b	6	6a	7	7a
0000	0	NUL		Sp		0			●		P		—		p	
0001	1		Cursor on	!		1		Set verify mode (1)	A	Alphanumeric Red	Q	Mosaic Red	a		q	
0010	2			"		2		Set verify mode (2)	B	Alphanumeric Green	R	Mosaic Green	b		r	
0011	3			£		3		Skip block	C	Alphanumeric Yellow	S	Mosaic Yellow	c		s	
0100	4		Cursor off	$		4		Set programme mode	D	Alphanumeric Blue	T	Mosaic Blue	d		t	
0101	5	ENQ		%		5			E	Alphanumeric Magenta	U	Mosaic Magenta	e		u	
0110	6			&		6			F	Alphanumeric Cyan	V	Mosaic Cyan	f		v	
0111	7			'		7			G	Alphanumeric White	W	Mosaic White	g		w	
1000	8	Active Position Backward (APB)		(		8			H	Flash	X	Conceal Display	h		x	
1001	9	Active Position Forward (APF)		)		9			I	Steady	Y	Contiguous Mosaics	i		y	
1010	10	Active Position Down (APD)		*		:			J		Z	Separated Mosaics	j		z	
1011	11	Active Position Up (APU)	ESC	+		;			K		—		k		¼	
1100	12	Clear Screen (CS)		,		<			L	Normal Height	½	Black Background	l		‖	
1101	13	Active Position Return (APR)		—		=			M	Double Height	—	New Background	m		¾	
1110	14		Active Position Home (APH)	.		>			N		↑	Hold Mosaics	n		÷	
1111	15			/		?			O		#	Release Mosaics	o		■	

Figure 4.3 Videotex code table

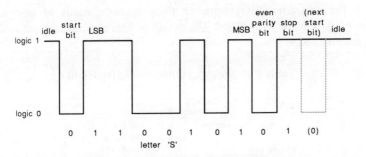

Figure 4.4   Videotex serial data format

4.4.1 Decoding the serial signal

At the receiving end, the signal is scanned for a logic 0 indicating a start bit, the signal is then tested again after a time interval corresponding to half the time for one bit and, if the signal is still at logic 0, then it can reasonably be assumed that the logic 0 is not due to interference or noise and is in fact a start bit. The signal is then tested at regular time intervals corresponding to the bit time, and the resulting logic levels should give the required 7 character bits, the parity bit and the stop bit. If the stop bit sample is at logic 1, and the parity check is correct, it is likely that the character has been correctly received.

Thus unlike teletext, the system relies on the extra start and stop bits to synchronise the decoder, and the decoder's timing circuitry must be sufficiently stable to allow sampling of the signal at the correct time during the nine bits following the start bit. In other words, the timing must not vary by more than half of a bit-time in nine, otherwise the last sample will not occur during the stop bit and an error will result.

4.4.2 Signal modulation

This logic signal cannot be sent directly down the telephone line, so it is modulated by frequency shift keying (FSK) which involves allocating one carrier frequency for the logic 1 level, and a different frequency for the logic 0 level.

The system can work simultaneously in both directions — that is, the operation is *full duplex* — and so in order that the two signals do not interfere, the carrier frequencies (and data rates) are different.

The actual values used are those of the V.23 communications standard, and in order to be able to use full duplex operation, the slow speed supervisory channel is used as the user-computer link, and the fast main channel for the computer-user link. This means that the data rate from the computer to user is fast, but the rate in the reverse direction is much slower. This situation is quite acceptable since the user is usually only keying slowly the page numbers required, whereas the computer, in response to these requests, needs to send a large volume of data back to the user. The details of this communications interface are as shown in table 4.1.

Table 4.1 Videotex communication details

Terminal to computer	
Carrier frequencies	logic 0 - 450 Hz ±5 Hz
	logic 1 - 390 Hz ±5 Hz
Data rate	75 bits/second ±1%
Computer to terminal	
Carrier frequencies	logic 0 - 2100 Hz ±16 Hz
	logic 1 - 1300 Hz ±16 Hz
Data rate	1200 bits/second ±0.1%

Notice that for the fast link, the FSK frequencies used are very close to the actual data rate — in fact, a single binary 1 results in only 1300/1200 = 1.08 cycles of 1300 Hz carrier being transmitted. This means that the demodulator in the decoder must be very carefully designed.

The speed used is of course limited by the carrier frequency, which in turn is limited by the bandwidth of the telephone lines — usually 3.4 kHz. 1200 baud is about the highest speed attainable without employing other, more expensive, methods of modulation.

4.4.3 Data rate

The average data rate will vary depending on the idle time between each set of data bits. However, the maximum data rate, where the idle time is zero, can easily be calculated, since for every character or control code transmitted, 10 data bits must be sent.

Hence, for the computer-user link, the data rate (D_{cu}) is

$$D_{cu} = \frac{1200}{10} = 120 \text{ characters/second} \tag{4.1}$$

and for the user-computer link the data rate (D_{uc}) is

$$D_{uc} = \frac{75}{10} = 7.5 \text{ characters/second} \tag{4.2}$$

Notice how slow the data rate for the user's page requests is — a good typist can easily type faster than this. However, with sufficient buffering, this speed is not too much of a problem except when large amounts of data have to be input to the videotex computer. For this reason, there is usually provision for a fast half duplex (one way at a time) or other fast communication method to input such data using special ports on the videotex computer.

5 Television Receivers

This chapter will look at the various sections of a typical television receiver — including those parts which are involved in teletext and videotex decoding and the control of the receiver. The chapter is not a detailed description on how to design television receivers; rather it uses a block diagram approach to illustrate the sort of basic functions that are performed inside all receivers.

A colour television receiver containing teletext and videotex decoders can be divided into a number of basic functional modules which are listed below:

1. Reception and demodulation of the television signal.

2. Demodulation and amplification of the sound signal.

3. Separation and processing of the synchronisation signals into a form suitable for driving the deflection coils of the picture tube.

4. Demodulation of the video signal into the separate red, green and blue colour signals and processing of these into a form suitable for driving the picture tube.

5. Separation and decoding of the teletext signal and creation of the red, green and blue colour signals.

6. Reception and decoding of the videotex signal and interfacing with the teletext decoder for generation of the display signals.

7. A control system operated by the user (by remote control or local buttons) to enable the user to operate the receiver.

8. A power supply to provide the various low and high voltages needed in the receiver circuitry.

Figure 5.1 shows a block diagram of a television receiver showing the interconnection of the above modules.

The operations being performed in each of these modules will now be considered in general terms. It will be assumed that the display device is a colour cathode ray tube such as that described in chapter 2 whose electron beam is deflected by magnetic scan coils.

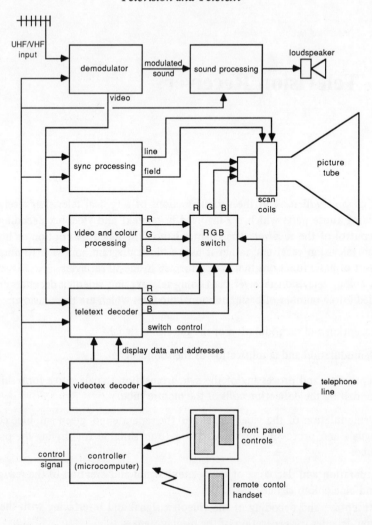

Figure 5.1 Block diagram of a typical television receiver

5.1 Reception and demodulation of the video signal

This section of the receiver involves taking the signal from the aerial and producing the composite video and modulated sound signals. Figure 5.2 shows a block diagram of the various sections of the module.

In order to recover the video signal from the UHF or VHF carrier successfully without interference from adjacent channels, close tolerance filtering is required. This can more easily be achieved at a fixed frequency. Hence television receivers

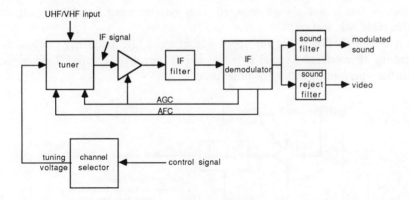

Figure 5.2   Demodulator module

use the superhet principle whereby the modulated signal is first down-converted to a lower, fixed, intermediate frequency signal (IF); the signal is then filtered carefully to leave only the wanted video signal, and finally this signal is demodulated to produce the unmodulated video signal.

5.1.1 *Tuner unit*

The aerial signal is first applied to a high-frequency, low-noise amplifier. Usually this amplifier is tuned to the rough band of frequencies containing the required video signal. Tuning is accomplished by applying a d.c. voltage to a *varicap diode* which has the property of changing its capacitance according to the applied voltage. Also, to prevent overloading of subsequent circuitry, the gain of this amplifier is controlled by a signal from the IF circuitry so that the amplitude of the resulting signal is substantially constant over the whole range of signal strengths.

The resulting signal is mixed with the output of a local oscillator whose frequency is exactly the IF frequency above or below the carrier frequency of the required video signal. The resulting lower sideband is thus the IF modulated signal. By changing the local oscillator frequency, the mixing process will produce other video channels whose carriers correspond to the fixed IF frequency. Thus selection of the required channel is achieved simply by varying the local oscilattor frequency (again by the use of a varicap diode). This section is therefore usually known as the *tuner unit*.

Use of the varicap diode to vary the oscillator frequency has the added advantage that channel selection can be achieved by a d.c. control signal. Also, since the frequency stability of this oscillator controls how well the receiver stays in tune to the required channel, an automatic frequency control (AFC) signal from the IF circuitry is added to this tuning voltage to keep the television in tune, although this is not always required with systems using frequency synthesis tuning (see next section).

After simple filtering of the upper sideband and high frequency carriers, the resulting IF modulated signal is fed to the IF circuitry. Figure 5.3 shows the functions performed within the tuner unit.

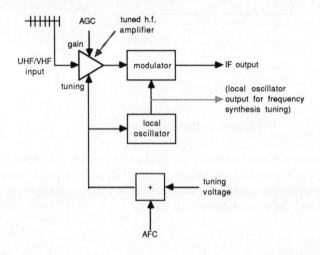

Figure 5.3 Tuner unit

5.1.2 Frequency synthesis tuning

Increasingly in modern receivers, the d.c. voltage used to tune the receiver to the required television broadcast is being provided as part of a phase locked loop; such a system is known as *frequency synthesis tuning*. This system has advantages for the reliability and stability of the receiver. Figure 5.4 shows a block diagram of a frequency synthesis tuning system in a television receiver.

The tuner unit's local oscillator frequency is fed to a high-speed divider (usually constructed using emitter coupled logic (ECL) technology) which reduces the frequency to a more manageable range. This frequency is then fed to a second, digitally programmable divider, which is arranged to produce a pre-defined frequency as long as the tuner local oscillator is running at the correct frequency for the required television channel.

The resulting frequency is compared with a crystal controlled oscillator operating at the pre-defined frequency. Any phase difference between the two frequencies is detected and used to produce a d.c. error voltage which can be suitably amplified to provide the tuner's tuning voltage.

Thus a control loop is created such that the tuner's local oscillator will remain as stable as the stability of the crystal controlled oscillator. This stability

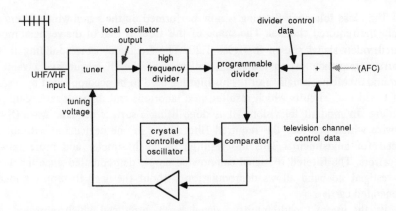

Figure 5.4 Frequency synthesis tuning

is such that no AFC signal from the IF circuitry is required – this means that the receiver will be more reliable since the production of the AFC signal relies on the use of *L-C* tuned circuits which are apt to vary with time.

Unfortunately, AFC is normally provided (but usually switchable) not because of the receiver, but because of the transmitter! The radio frequency carrier produced from video cassette recorders, video games etc., and even some local sub-transmitter stations, is not sufficiently stable and AFC is provided to enable the receiver to 'track' the transmitter carrier. However, main broadcast transmitters are very stable, and during reception of these channels the AFC can be switched off.

Tuning of the receiver is therefore accomplished by altering the divide ratio of the variable divider. This can be accomplished digitally by the user interface control unit. Thus by using a simple algorithm, the control unit can convert actual UHF or VHF channel numbers supplied by the user or stored in the receiver to the divide ratio needed to tune the receiver directly to the required television station. The control unit can also switch the AFC signal on or off when required.

The advantage of this system for the consumer is that tuning is easy, and the receiver will not drift out of tune with time (many service calls are simply because receivers have drifted out of tune and the user is unable to retune the set). For the manufacturer, the advantage is that the television can be installed and tuned to the required channels even if they are not 'on the air' at the time, although the disadvantage is that there is a small extra production cost.

5.1.3 *IF signal processing*

The IF signal from the tuner unit is first amplified by an amplifier whose gain is also regulated by the IF gain control signal to produce a constant-amplitude IF signal.

The close tolerance filtering is now performed on the signal which filters out only the required channel. The shape of this filter is that of the vestigial modulated video signal shown in figure 1.20. The frequency corresponding to the carrier frequency on that diagram will now be the IF frequency (a value of around 39 MHz is usually used). This filtering used to be accomplished by a series of tuned L-C circuits which resulted in a laborious and difficult setting-up procedure. In modern televisions, it is done using a surface acoustic wave (SAW) device which provides the required filter shape in one device and without the need for adjustments. This has resulted in much simpler and more reliable receivers. The filtered IF signal can now be simply demodulated since the use of a vestigial sideband allows demodulation without the need to supply a locally generated carrier.

By the use of tuned circuits, a signal can be produced which represents any phase difference between the required IF frequency and the actual frequency. This signal can be fed back as the AFC signal to control the tuner unit's local oscillator, as mentioned above. Also, by using a line flyback signal produced in the synchronising circuitry, the level of the modulated signal during the line sync pulse can be monitored and any variations can be used to provide the automatic gain control (AGC) to the tuned IF and tuner unit amplifiers. Since the IF signal amplitude during the line sync pulses should be constant, this signal will ensure that the resulting video signal will not vary in amplitude under differing conditions of reception.

Finally, the video signal is split two ways — one to be filtered to reject the sound carrier and leave just the composite video signal, and the other to filter out only the sound carrier. This signal can then be demodulated and amplified to produce the sound signal as explained below.

In most of today's receivers (1988) much of the IF processing is achieved in a single integrated circuit with only a few external components.

5.2 Audio signal processing

In this module, the sound signal is demodulated, amplified and fed to the receiver's loudspeaker.

The details of the demodulation will be different for each particular system in use. However standard techniques are used to demodulate either the FM or AM signal from the IF demodulator and filter mentioned in the previous section. In the case of an FM signal, de-emphasis is then applied to the resulting audio signal.

After demodulation, the signal is fed to the audio power amplifier and then on to the receiver's loudspeaker. Volume control is usually achieved by a d.c. control rather than a mechanical potentiometer since this allows the possibility of control by a microcomputer control system.

In a receiver with a videotex decoder, there will be a sound input from the decoder after the volume control. This allows monitoring of the telephone line during set-up of a videotex session — the volume control being ineffective in this case, as required by the original videotex specification.

Figure 5.5 shows a block diagram of the audio section of a television.

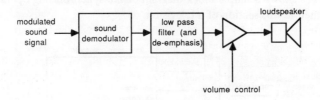

Figure 5.5 Sound processing

Like the IF processing, sound demodulation is now usually carried out inside a single integrated circuit — which also usually contains a pre-amplifier and the d.c. volume control circuitry.

5.3 Synchronisation signal processing

The purpose of this module is to recover the sync pulses from the composite video signal and process them to provide suitable signals to drive the deflection coils of the picture tube.

This section of a receiver is probably the most complex, although the ideas are very simple. The problem is mainly in how to shape the line and field scan waveforms in order to produce a rectangular picture (rather than a picture that is convex or concave along its top or sides).

At first thought it would seem that a simple sawtooth waveform as shown in figure 5.6 applied to the picture tube scan coils would produce a rectangular picture. However this is not the case, and although the basic scan waveforms are sawtooth, some additional processing has to be applied to generate the required scan signal.

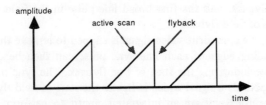

Figure 5.6 Linear sawtooth scanning waveform

The detail design of the scan circuitry is thus fairly complex, and this section will only look generally at the functions that have to be performed. In fact, the actual circuit of a typical sync processing section in a television would not appear as the separate circuit blocks to be described below — although all of the functions described are incorporated at some point.

Figure 5.7 shows a simple block diagram of the synchronising operations that are performed in this section of a television.

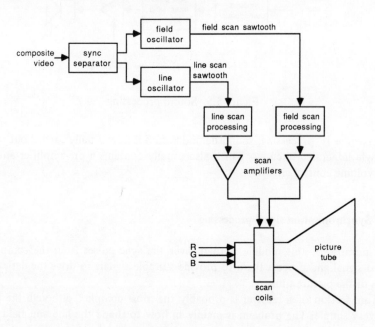

Figure 5.7 Simple diagram of the sync processing circuitry

5.3.1 Sync separator

This section produces line and field flyback pulses from the composite video waveform. Remember that the leading edge of the line sync pulse indicates the start of line flyback, and the first broad line pulse in the field sync pulses indicates the start of field flyback.

First of all, the composite video signal is clipped to remove the video information. The leading edge of each line sync pulse can then be differentiated to provide a pulse indicating the start of line flyback. The sync pulse train is also integrated to provide a signal which indicates the start of field flyback.

Modern receivers employ an integrated circuit to perform these functions, and digital techniques are also used within these devices to provide the line and field flyback pulses.

An additional function of this section is to provide line and field blanking signals to indicate when the video signal should be switched off during line and field flyback, and a signal for the colour decoder (NTSC and PAL systems) indicating the position of the colour sub-carrier burst. This signal is usually combined with the line blanking signal and is called the *sandcastle pulse* as illustrated in figure 5.8.

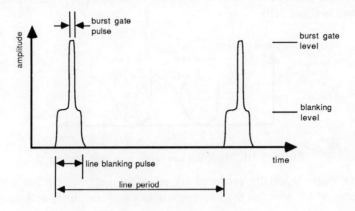

Figure 5.8 Sandcastle pulse

5.3.2 *Line oscillator and waveform shaping*

To ensure that the scanning motions of the picture tube electron beam never stop, even when there is no video signal being received, a sawtooth oscillator constantly runs at a rate a little less than the normal line scan time to provide a scan signal at all times. When a video signal is being received, the resulting line flyback pulses are fed to the oscillator to initiate the start of line flyback. Thus the line scan oscillator is locked to the incoming video signal. This process is called *flywheel synchronisation*.

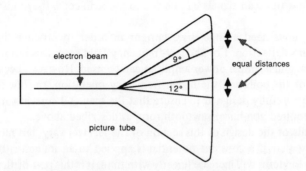

Figure 5.9 Deflection angles to produce equal scan distances

As explained above, a linear sawtooth waveform will not provide a rectangular picture. This is because the angular deflection of the electron beam is roughly proportional to the applied scan coil voltage rather than the linear deflection across the screen. This is illustrated in figure 5.9 which shows that the deflection angle required to produce the same linear scan distance on the screen is different at different points across (and down) the screen.

Thus the linear sawtooth waveform must be modified, and the required shape is shown in figure 5.10.

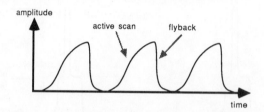

Figure 5.10   Modified sawtooth scanning waveform

This shape is usually produced by an additional circuit called a *parabola generator* which modifies the linear sawtooth waveform during each active scan.

5.3.3 *Field oscillator and waveform shaping*

The same process as for the line scan is required for the field scan signal — although the frequency of the field scan oscillator is of course much lower.

Thus a free running field scan oscillator is locked to the incoming video signal by the field flyback pulses initiating the end of a field scan, and the linear field sawtooth is modified to produce the required non-linear signal to drive the vertical deflection coils on the picture tube.

5.3.4 *Scan signal amplifiers*

Finally, before the scan signals can be fed to the scan coils, they must be amplified.

The scan coils need a large drive current in order to produce the required electron beam deflection. For the low-frequency field scan, this is not too much of a problem, and a simple power amplifier can be used. However, because of the inductance of the coils and the higher-frequency operation, the line scan amplifier must be carefully designed to ensure that the scan coil current varies according to the required non-linear sawtooth form as described above.

The details of the design of this section of the receiver vary, but most employ the property that if a constant potential is applied to an inductor, the current through the inductor will increase linearly with time. It is this part of the receiver's circuitry which has evolved the cleverest designs, although they are beyond the scope of this book.

Television Receivers

In both the line and field amplification sections, provision must be made for adjustment of both the line and field position and scan width to enable the picture to be correctly positioned on the screen.

5.4 Video signal processing

This section of the receiver is concerned with producing the three colour signals — red, green, and blue — from the composite video signal in a form suitable for driving the picture tube. Figure 5.11 shows the basic operations that this module performs (for all colour encoding systems).

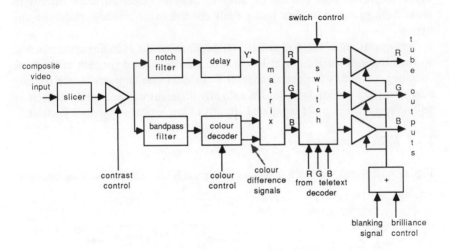

Figure 5.11 Video signal processing

The composite video signal is first processed to remove the sync information. This leaves the monochrome signal (Y') and the chroma signal.

A controlled amplifier then provides the user's contrast control (that is, the amplitude of the video signal) and the resulting signal is split two ways. Firstly it is filtered to remove the chroma signal (this is the notch filter mentioned in chapter 2), thus producing the original Y' monochrome signal.

Secondly, the signal is filtered to separate out only the chroma signal. This is a bandpass filter designed to have a bandwidth equal to the chroma bandwidth of the particular colour encoding system in use.

The Y' signal is then delayed to compensate for the delay introduced by low pass filters in the colour decoder and fed to the matrix circuit where it is combined with the two colour difference signals in the correct ratios to produce the three primary colour signals.

In a receiver that contains a teletext or videotex decoder, there is then a switch circuit which allows either the picture video signals or the text signals from the text decoder(s) to be fed to the picture tube. The control signal for this also comes from the text decoder and is a fast signal which can operate at video frequencies to enable subtitles etc. to be inserted into a television picture.

Finally the resulting red, green and blue signals (text and/or picture) are amplified and usually fed directly to the cathodes of the three electron beams in the picture tube. These amplifiers have a control input to allow the user to adjust the brightness of the final picture (that is, the absolute d.c. level of the video signals). Also, the blanking signal from the sync processing module is fed to the amplifiers to cut off the drive to the tube during line and field flyback. These amplifiers also contain circuitry to limit the electron beam current to avoid burning the tube face under fault conditions or mal-adjustment by the user.

The operation of the colour decoder depends on the encoding system in use, so the functions carried out in this section will be studied for each of the three encoding methods described in chapter 2. In each case, a practical receiver may implement the various operations in a slightly different way and modern receivers will now use one or more integrated circuits to implement a complete decoder.

5.4.1 NTSC colour decoder

Figure 5.12 shows the block diagram of a typical NTSC colour decoding circuitry.

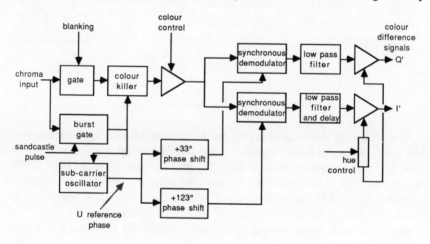

Figure 5.12   NTSC colour decoder

The chroma input signal is first gated to remove the burst, and also gated to provide a second signal containing only the burst. The control signal for these gates is the sandcastle pulse produced in the synchronisation module.

The gated chroma signal is then fed through a second gate which disconnects the output when there is no or insufficient burst amplitude. This acts as a colour killer to revert the picture to monochrome in the event of there being no subcarrier burst present in the transmission. A controlled amplifier then allows the user to set the amount of colour in the final picture by adjusting the chroma amplitude.

Meanwhile, the gated burst is also used to phase-lock a crystal controlled subcarrier oscillator whose output is phase shifted by +33° and +123° to provide the Q' and I' signal reference phases respectively.

Two synchronous demodulators then demodulate the chroma signal to produce the separate Q' and I' chrominance signals. Low pass filters then filter out unwanted high-frequency components, and the reduced bandwidth of the Q' signal means that an extra delay has to be inserted in the I' signal path to compensate for the reduced filter bandwidth.

The signals are finally amplified by controlled amplifiers which allow the balance of the two chrominance signals to be adjusted either internally or by the user (this is the 'hue' control found on some receivers which can be used to compensate for static phase errors).

5.4.2 PAL colour decoder

Figure 5.13 shows the block diagram of a typical PAL colour decoder.

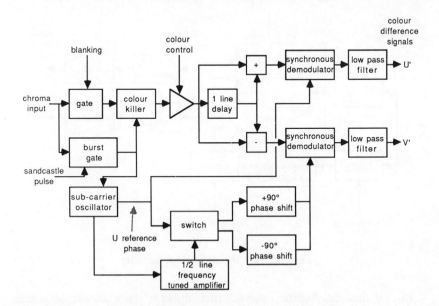

Figure 5.13 PAL colour decoder

As can be seen, most of the functions performed in the PAL decoder are identical to those in the NTSC decoder described above. The difference lies in the demodulation of the chroma signal.

To average out the effects of phase errors, the chroma signal, after controlled amplification, is delayed by an exact one line scan time. Undelayed and delayed signals are added and subtracted to produce the separate U' and V' chroma signals ready for demodulation (see chapter 2).

The reference phase sub-carrier oscillator output is used to demodulate the U' signal directly. The V' demodulator is however fed with sub-carrier phases of +90° and −90° on alternate lines. This provides the 'V switch' to ensure the V chrominance signal is inverted every line as required (chapter 2). The 'V switch' is controlled by the output of a one half line frequency tuned amplifier which is fed from the sub-carrier oscillator. The 90° phase change of the burst produces a half line frequency signal at the amplifier output which can be used to effect the V' sub-carrier phase inversion on alternate lines.

The demodulated U' and V' signals are then filtered and amplified but, unlike the NTSC decoder, there is no need for an extra delay in one of the chrominance signal paths since both signals have the same bandwidth. Also, there is no need for a hue control to balance the chrominance signal amplitudes.

5.4.3 SECAM colour decoder

Figure 5.14 shows the block diagram for a typical SECAM colour decoder.

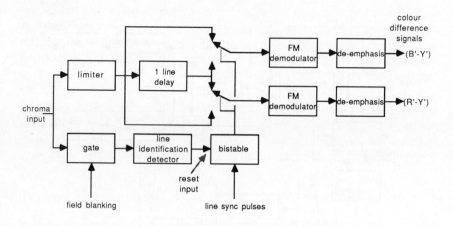

Figure 5.14   SECAM colour decoder

The FM modulated chroma signal is first limited to give a constant-amplitude signal. Remember that the transmitted chroma amplitude increases for increasing chrominance amplitude to reduce chroma effects on monochrome receivers. This is introduced purely to reduce these interference effects and is not relevant to the demodulation process.

The limited chroma signal is then fed to a one line period delay, and two switches select either the delayed or undelayed signal on alternate lines. The D'_R and D'_B signals are thus provided on every line.

These signals are then FM demodulated and filtered. Before final amplification, de-emphasis is applied to cancel the effect of the low-frequency pre-emphasis applied to the signal when it was encoded.

The selector switch mentioned above is operated by a bistable circuit which is clocked by line sync pulses from the synchronising module. To ensure the bistable is in the correct state on each line, the chroma signal in the field blanking interval is fed to an identification circuit which resets the bistable to the correct state if indicated by the SECAM identification lines contained in the field blanking interval. Thus there is a maximum period of one field scan with the chrominance signals reversed after the receiver is switched on, or a channel is changed, or any other interruption of the video signal occurs.

5.5 Teletext decoder

This module's functions are to extract and decode the teletext data from the composite video waveform and to provide the three colour drive signals to enable the teletext data to be displayed.

The decoder is complicated and contains a combination of analogue and digital circuits. Although in the early days a few were made using discrete components, nowadays all decoders use large scale integrated circuits. The first generation of these required about four devices, but complete teletext decoders have now been integrated on to two devices — one to perform the analogue functions and the other for the digital functions.

In the following description, an outline of the operations a teletext decoder has to perform will be given. No attempt will be made to describe the detailed operation of the various analgoue and logic functions contained within each section of the decoder — such detail would require a book in itself.

Figure 5.15 shows a block diagram of the main sections of a teletext decoder.

5.5.1 Sync and timing generator

The incoming composite video is first split into the sync and video components. The sync signal is fed to a local sync and timing generator and causes the local sync generator to lock in phase with the incoming video.

This circuit has two functions:

(a) It provides a local sync signal for the receiver so that in the event of the television signal being removed, a stable teletext display will still be available

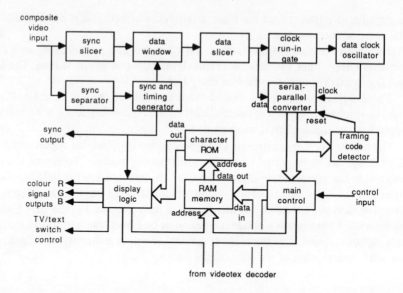

Figure 5.15 Teletext decoder

(this is vital for videotex, since there does not need to be a television signal to use a videotex system). This circuit provides a combined line and field sync signal that can be fed to the synchronisation module described above rather than from the actual received television signal.
(b) It provides all the necessary timing signals required by the rest of the teletext decoder. These include signals such as a data enable window to indicate when teletext television lines may be expected (that is, those lines in the field blanking period after the field sync and equalising pulses); a signal to indicate the clock run-in period at the start of each teletext television line; and fast dot clock signals for the display section of the decoder.

5.5.2 Data signal conditioning

The separated video signal is fed to two circuits that condition it into a digital form. The first is a gate which only allows television lines which might contain teletext data to pass — as described in (b) above. The second is a data 'slicer' that produces clean logic levels from the video signal (the video signal itself contains raised cosine pulses and is probably reduced in level and distorted by the transmission path).

The data slicer involves a great deal of clever design to enable the best performance to be obtained in the presence of a distorted or noisy signal, and especially in the presence of multipath reflections. Usually a signal level is set, above which a logic 1 and below which a logic zero is produced. The switching level is varied according to the average signal level (both in terms of absolute d.c. level and signal amplitude) — such a device is termed an *adaptive data slicer*.

This section is the only part of the decoder that contains analogue circuitry, all the remainder requires solely digital techniques.

5.5.3 Data clock generation

The next stage in the decoding process is to generate a local clock oscillator which can be used to sample the logic data from the data slicer to determine the state of each bit. Remember that the teletext data proper does not contain any clock information, this is contained in the first two data bytes — the clock run-in.

The period where the clock run-in might be expected (that is, the 16-bit period after the end of the line sync pulse on a possible teletext television line) is gated out and the clock run-in (if present) used to lock a local clock oscillator in phase. This clock is sufficiently stable to enable the rest of the data on that line to be sampled at the correct instant in order to be correctly decoded.

5.5.4 Serial-to-parallel conversion

The serial data is actually sampled by being clocked into a serial to 8-bit parallel shift register using the local clock oscillator. A framing code detector scans the parallel data looking for the unique framing code which occurs after the clock run-in.

When (and if) the framing code detector finds a correct framing code, it clears the serial-to-parallel converter and sends a signal to the main control unit to indicate that a valid teletext packet has been detected, and bit and byte synchronisation has been achieved.

The rest of the data in the teletext packet is then shifted into the serial-to-parallel converter and the control unit reads each data byte as it becomes available.

5.5.5 Main control logic

The overall function of this section is to control the acquisition of the required teletext page and to store the data in the display memory ready for display.

The control unit contains a set of latches which are set via the control input from the user to contain the magazine, page and sub-code numbers of the required page. When a new page is selected, these latches are updated and the control unit searches for a header packet with the required page numbers. While this process is being carried out, most decoders will directly store in the display memory all incoming header rows. This has the effect of producing a 'rolling' header display so that the user can see the cycle of page numbers being transmitted.

When the header of the required page is detected, the rolling header display is stopped, and the display memory is cleared (all locations are set to the space code (2/0)). Subsequent teletext packets with the same magazine number are then stored in the display memory in the correct position according to their row numbers. When the next header in the same magazine is received, this process is stopped and the page capture is deemed to have been completed.

On subsequent detection of the required page (that is, when it is transmitted during subsequent broadcasts of the magazine) the control unit does not clear the page, but does update the display memory. This has the effect of (a) showing any updates to the information contained in the page, and (b) writing characters to the display which in the previous cycle were not written because of a parity error.

If the control bits in the header of the required page indicate the page is a subtitle or newsflash page, then the control unit sets the red, green and blue colour signal switch control to display the television picture, and the display unit then controls the insertion of text into this picture on receipt of the required start/end box display control codes.

The control unit therefore controls the overall acquisition of pages. It takes command inputs from the user, and controls the writing of teletext data into the display memory.

5.5.6 Display memory

The display memory is a standard RAM memory which is large enough to hold the data for one page. Since there are 40 × 24 character locations on a page, a standard 1K byte memory is sufficient to store the data for a single page. Writing into the memory is controlled by the control unit as mentioned above, and the data is read by the display unit during the active part of each scan line so that the data can be converted into the shapes of the characters and displayed on the screen.

For normal teletext, there is no conflict between writing and reading, since data is only written during the field flyback interval when data is not required from the memory for display purposes and vice versa. However for full channel operation, some form of multiplexing has to be incorporated since data has to be stored at the same time as data is being read for display. This may be achieved by buffering the data line by line either in the display unit or control unit or both, or by synchronising the memory read and write cycles so that they do not occur simultaneously.

5.5.7 Display unit

The final unit in the teletext decoder takes the data codes in the display memory, and from them produces the red, green and blue video signals which form the picture of the data.

The operation of the display unit is controlled by sync signals from the timing unit. The data is read sequentially from the display memory, and each code (if it is not a display control code) is used to address an internal character ROM which forms the actual shape of the required character. The output of this ROM then forms the three colour video signals after being gated by various gates set by the serial control codes so as to produce the required display colour and attribute. These gates are reset at the start of each scan line so that in the absence of control codes, a white, steady, normal height display is produced on a black background.

If graphics characters are required, the serial graphic control code sets a latch so that another part of the character ROM containing the graphic character forms is accessed.

The display unit also produces a fast switching output to enable characters to be boxed in the television picture. This is combined with the overall picture/text signal from the control unit and is fed to the video switch mentioned earlier.

5.6 Videotex decoder

The videotex decoder receives page request commands from the user, converts them into the required ASCII format and transmits them down the telephone line. It also receives data from the videotex computer, and places it in the display memory ready for display by a display controller. The decoder may also be able to perform auto-dial and remote user/telephone number programming functions.

If the television receiver also contains a teletext decoder, the videotex decoder's display data will be stored directly in the teletext display memory to be displayed by the teletext decoder's display controller. This is because the display formats of teletext and videotex are identical, thus the display memory and controller do not have to be duplicated in a receiver with both types of decoder.

The main components of a videotex decoder are as shown in figure 5.16.

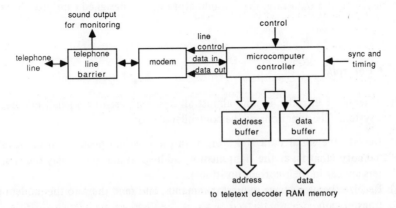

Figure 5.16 Videotex decoder

5.6.1 Telephone line barrier

The telephone line barrier provides the electrical isolation between the telephone line and the television receiver. This isolation is most important for the safety of both the television receiver and the public telephone system.

The line barrier prevents surges on the telephone line (from lightning etc.) from damaging the receiver, and it also prevents any possibility of the very high voltages occurring in the picture tube drive circuitry from getting into the telephone system and causing a potential hazard. For these reasons, the line barrier (and the whole modem as well) has to be approved by the telecommunications authority before it can be installed inside a television receiver.

5.6.2 Modem

The modem is the unit which modulates and demodulates the outgoing/incoming data.

For outgoing data, the serial binary codes from the controller are used to alter the frequency of an oscillator between the two transmit frequencies (450 Hz for binary 0 and 390 Hz for binary 1). The resulting FSK signal is sent to the line barrier for transmission.

The incoming signal is demodulated as follows. The signal is first filtered and limited to give a set of rectangular pulses. The time between each zero crossing of the resulting waveform is then measured and the resulting times used to determine whether a binary 0 or binary 1 is being transmitted. Since a binary 0 is modulated using the higher carrier frequency, and also the 'no signal' condition is indicated by a steady binary 1 (lower-frequency tone), the modem is simply set such that if the time between zero crossings is less than a pre-determined interval, then a binary 0 is output, otherwise a binary 1 is output.

This sounds very simple, but because of the nearness of the lower carrier frequency to the data rate, the demodulator must be designed carefully to avoid errors.

5.6.3 Controller

The videotex controller is now almost always composed of a small microcomputer system. The functions of the controller are to:

(a) Decode the serial data and store it in the required position in the display memory (that is, at the right memory address so that on display the characters appear in their correct positions).
(b) Receive the user's page request commands, and send them to the modem for transmission.
(c) Perform the auto-dial and programming functions.

(d) Interface with the display memory and controller so that there is no conflict over memory access for data storage and retrieval for display.

Suitable programming of the control computer can achieve these functions without much difficulty. However since the system is operating in real time, some careful planning of the program must be carried out to ensure that either data is not lost or that the display memory is accessed when the display controller is reading the memory. If the computer is sufficiently fast and powerful, parallel/serial and serial/parallel conversion of the data can be carried out by software without using a UART device, so resulting in a cheaper decoder.

To prevent conflicts over access to the display memory, the controller is supplied with sync signals. The display controller will require access to the display memory during the active part of each scan line; at other times the memory will be available for updating with new display data — assuming teletext acquisition is inhibited. The controller therefore updates the display memory either during any line blanking period, or during the field flyback interval. This may require the computer to have its own small section of RAM memory for buffering purposes.

5.6.4 Display memory and controller

These have identical functions to those in the teletext decoder, and cause the data to be displayed on the television screen. As mentioned above, for a receiver with both a teletext and a videotex decoder, only one display memory and controller are required.

5.7 Control module

In every television receiver there must be a control unit to enable the user to set the various controls.

In a simple receiver, without teletext or videotex decoders, the control unit may consist solely of a set of potentiometers mounted on the front panel, which supply the d.c. voltages to set the main television controls — brilliance, contrast, colour and volume — and a second set of potentiometers and a selector switch to provide the tuning voltage to tune in the receiver to each television channel. Such a control unit is illustrated in figure 5.17.

In a more complicated receiver with teletext and/or videotex decoders, a more sophisticated control system must be incorporated.

Receivers with these facilities usually have remote control as well as front panel controls, and the various units within the receiver are controlled by a common bus signal. This principle is shown in figure 5.18 which shows the block diagram of a control module for a remote control teletext/videotex receiver.

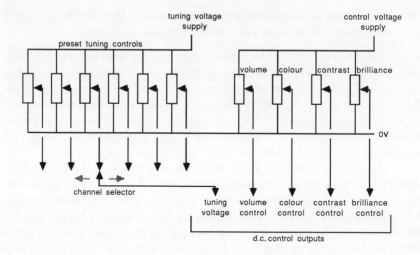

Figure 5.17 Simple control unit

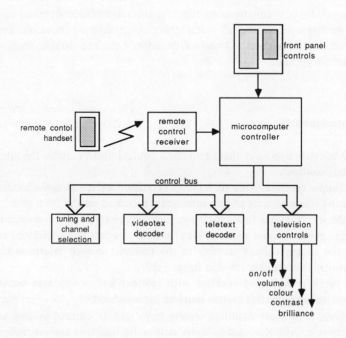

Figure 5.18 Sophisticated control unit

5.7.1 Remote control transmitter

This device transmits a control signal whenever a button is pressed. Early devices used ultrasound, but infra-red light is more commonly used now. The device contains circuitry which detects a button push, and activates its transmitting section only when required. This means that there is no need for an on/off switch on the device since the current drain on the batteries is very low until a button is actually pressed.

The signal transmitted is in the form of a serial control code, which is usually transmitted continuously while the button is held down.

5.7.2 Remote control receiver

This contains a detector (microphone for ultrasound systems or light-sensitive device for infra-red systems) and a gain controlled amplifier that produces as far as possible a clean serial binary signal representing the code transmitted. The output of this device is fed to the main controller for decoding.

5.7.3 Front panel controls

The front panel controls usually consist of a set of buttons which duplicate some or all of the remote control functions. Quite often this unit contains a remote control transmitting device, but without the final amplifier and transducer – the binary output from the device being fed directly to the main control unit.

5.7.4 Main control unit

This device is the heart of the control module, it receives and decodes commands from the local and remote transmitters, and supplies signals along the control bus that operate the various controls.

In the early days of remote control, this unit contained a dedicated integrated circuit to perform all the functions. This had the disadvantage that it was not flexible, and television manufacturers were not able to modify the mode of operation of the unit to suit their own particular requirements.

In modern televisions a dedicated microcomputer is used to perform these control functions. The device is usually a mask programmed computer, but the advantage is that the operation of the control unit can be determined by the television manufacturer, not by the integrated circuit manufacturer.

This computer may also have some non-volatile RAM as well to enable control settings to be retained when the receiver is switched off. These settings may include the states of the various analogue controls (brilliance, colour etc.), initial teletext page numbers to capture on each television channel as well as the frequency synthesis divide ratios to permit tuning to the various television broadcasts.

5.7.5 Control bus

The control bus is the nerve centre of the receiver. Signals are transmitted down the bus to all the modules requiring control by the user.

One of the most common of these buses is the *inter-IC bus* originally developed in Britain by Mullard – a subsidiary of Philips – but now adopted by other manufacturers as well. This bus, usually abbreviated to I^2C (I squared C) uses a synchronous two-wire plus earth system. Data is set down one wire, and clock information is transmitted on the other.

The I^2C bus provides for any number of originators and receivers. The bus protocol contains an arbitration method to prevent more than one orginator transmitting data simultaneously. In the television control system, there is however usually only one originator – the control computer – all the other devices on the bus being receivers.

The data on the bus starts with a device address to identify which device the data is intended for, then a sub-address to indicate the function within the device that is to be altered, and finally the data proper, which is of indeterminate length. A stop condition indicates the completion of a bus transfer, leaving the bus free for further transfers.

Such bus systems are becoming common in television receivers as the functions requiring control become more complicated.

5.7.6 Bus receivers

The units connected to the bus are the actual devices requiring control. These are the teletext and videotex decoders, and special devices that interface to the control bus and provide the analogue voltage outputs for the various basic controls and channel tuning. These devices can be mounted within the modules they control, giving simplified wiring between modules.

5.8 Power supply

The final module in the receiver is the power supply. The problem with the power supply is the wide range of voltages and currents required – from 5 V for the logic circuitry to 25 kV for the picture tube final anode.

Originally, to save cost, the supplies were formed by directly rectifying the mains supply – the neutral of the supply being connected to the receiver's metal chassis. This rectified mains supply was used to drive the line scan drive circuitry, which contained a line scan transformer operating at line frequency to provide the remainder of the supplies. Even with the advent of transistorised circuitry, a mains transformer was not used. This means that the television had to be constructed very carefully so that there could be no possibility of any metal parts being exposed to the user. This became a great disadvantage when add-on units such as video tape recorders, separate text decoders, computers etc. were required. The only way they could be interfaced to the receiver was by the aerial socket. This resulted in a lowering of the quality of the picture.

Modern receivers now use a mains transformer to provide isolation between the receiver and the mains supply. However, again to save cost, the mains transformer is sometimes driven at a high frequency by a switched mode power supply (which must be isolated from the rest of the circuitry). This means that a much smaller, lighter and cheaper transformer can be used compared with one operating at mains frequencies.

5.9 Periconnector

The result of this isolation of the receiver from the mains supply is that peripheral devices can be freely connected. There is now a specification for an all-purpose interface connector called the *periconnector* which manufacturers are starting to include on their televisions. This connector contains connections for video and sync in and out; red, green and blue picture tube drive inputs; sound in and out; control in and out. Thus the television can be used in a whole host of new applications using this single connector.

Figure 5.19 shows the layout of the periconnector plug and table 5.1 details the functions of each of the connector pins.

Figure 5.19   Periconnector plug layout

5.10 Summary

This concludes a brief look at the circuitry inside a television. As explained at the start, this chapter has not attempted to show how to design a television, but has used a block approach to show in general terms the functions needed inside a receiver to enable the pictures, text and sound to be presented to the user.

Many of the units described are now available as single integrated circuits — in fact a complete basic television (without text decoders) can be constructed from a few ICs and a little discrete circuitry for the tube drive circuits and power supply.

Table 5.1 Periconnector pin functions

Pin number	Function	Input/output	Electrical details
1	Audio output B (stereo channel Right)	O	0.5 V r.m.s. nominal
2	Audio input B (stereo channel Right)	I	0.5 V r.m.s. nominal
3	Audio output A (stereo channel Left)	O	0.5 V r.m.s. nominal
4	Audio common return		
5	Blue signal return		
6	Audio input A (stereo channel Left)	I	0.5 V r.m.s. nominal
7	Blue signal	I	0.7 V / 75 Ω blanking to peak white levels
8	Picture & sound source selector	I	Logic 0 (0 to +2 V)-broadcast signals Logic 1 (+9.5 V to +12 V)-periconnector signals
9	Green signal return		
10	Control line 2		Presently unspecified
11	Green signal	I	As pin 7
12	Control line 1		Presently unspecified
13	Red signal return		
14	Control lines return		
15	Red signal	I	As pin 7
16	Blanking signal	I	Logic 0 (0 to +0.4 V)-picture on Logic 1 (+1 V to +3 V)-picture blanked Operates at video bandwidth
17	Video return		
18	Blanking return		
19	Composite video output	O	1 V / 75 Ω
20	Composite video input	I	1 V / 75 Ω

With the advent of the periconnector and text decoders, the control aspect of the television is becoming more important, and much more emphasis is being placed on the design of this section to achieve as much user-friendliness as possible.

6 New Specifications

This final chapter will look briefly at some of the new specifications that are being developed in the field of television and teletext. Like all aspects of electronics, the technology is changing rapidly, and this has enabled new specifications to be developed to provide enhanced services by employing more sophisticated techniques.

Earlier chapters have described the basic fundamentals of television and teletext which will be applicable as long as the systems remain in their present form. Some of the new specifications, however, involve completely new ideas and techniques and are not simply enhancements of the present systems.

First, two evolving aspects of the transmission of the television picture will be described − that of stereo digital sound transmissions, and the MAC specification devised primarily for transmission of television via satellites. The final two sections will look at developments occurring in teletext and videotex which allow a whole range of extra features and facilities to be used.

Space does not allow a detailed study − indeed a complete book would be required to describe the details of each of the specifications. It is hoped, however, that the descriptions will give a sufficient introduction to show in general terms what they encompass.

6.1 Stereo digital sound

This section looks at the specification adopted for the transmission of stereo sound in Britain. The method is now being adopted by other countries.

6.1.1 Choice of method

Until recently, the emphasis in television receiver development has been on improving the picture quality, the quality of sound being not considered so important.

The broadcasters, however, have been transmitting high-quality FM modulated sound ever since the high definition 625 and 525-line systems were introduced.

113

Such quality has not generally been realised because of the poor quality amplification and small speakers found in many television receivers.

With the advent of video recorders capable of reproducing stereo sound and of the developments in digital sound recording, engineers studied the possibility of modifying the composite video waveform to incorporate high-quality stereo sound.

The method of achieving this that immediately springs to mind is to use the AM sub-carrier and pilot tone system similar to that used in FM stereo radio broadcasts. There are however four other alternatives that were considered:

(a) Using an additional low-frequency FM sub-carrier modulated with the existing sound channel.
(b) Using bursts of digital data in the sync pulses (known as *sound in syncs*).
(c) Using an additional high-frequency sub-carrier.
(d) Modulation of an additional sub-carrier by a digital signal.

The first consideration is to ensure the bandwidth of the resulting composite signal is not increased such that (a) existing distribution channels are unable to carry the modified signal, and (b) the new signal affects programmes on adjacent radio frequency channels. The second is to ensure compatibility – existing receivers must be able to reproduce high-quality monophonic sound.

It was not possible to find any suitable space within the existing composite video bandwidth, hence if an extra signal were to be incorporated it would have to extend the bandwidth. Studies have shown that it is possible to increase the bandwidth by a small amount without causing the problems mentioned above, as long as the frequency components of the signal above the existing sound carrier are at a low level.

Having ascertained that bandwidth is available, the next consideration is whether to use an analogue or digital method of transmission. An analogue system based around either the AM sub-carrier plus pilot tone system or method (a) or (c) above has the advantage that the resulting signal will be both compatible and reverse compatible. However, developments in the digital reproduction of sound, as well as the specification of the MAC system for satellite broadcasting, resulted in a digital method being adopted. After many experiments of the two digital methods mentioned above, method (d) was finally decided upon, partly because of its compatibility with the MAC system. This does mean that the signal will not be reverse compatible – thus receivers designed to decode the stereo sound will have to incorporate a conventional sound demodulator for use on channels where there is no digital sound.

The method adopted and its effect on the bandwidth of the composite video signal will now be studied.

6.1.2 Analogue-to-digital conversion

The sound signals are first sampled at a rate of 32 kHz. The samples are then quantised into a 14-bit digital signal; this gives ample dynamic range for the audio signal under domestic listening conditions. Unfortunately there is insufficient bandwidth available in the video signal to allow all 14 bits to be transmitted, hence the signal is compressed to 10 bits using a form of near instantaneous companding. A single parity bit is added to these bits to give limited error protection.

The companding process is achieved by dividing the 14-bit samples into blocks of thirty two (a 1 ms time interval). All of the samples within each block are coded using a 10-bit two's complement code formed from the top ten significant bits of the sample with the largest amplitude within the block; for example if the largest amplitude sample uses only the lowest 10 bits, then the top 4 bits will not be used, if however the largest sample uses the most significant bit, then the top 10 bits will be used and the lower 4 bits will be discarded. A parity bit is added to these 10 bits to form an 11-bit serial code and the overall rate of the data for each audio channel is thus 11 x 32 000 = 352 kbits/second.

A 3-bit scale factor is also transmitted to indicate the degree of compression of the block. These bits are transmitted by modifying the parity bits of the samples in a pre-defined way. The precise details of the method are not important, but basically each parity bit is exclusive OR-ed with one of the three scale factor bits. In the receiver, the scale factor is recovered by performing the reverse process and taking the majority result as the correct scale factor bit. This means that the parity bits act as both error protection and as a means of transmitting the scale factor without the need for extra bits.

Table 6.1 summarises the characteristics of the conversion process.

Table 6.1   A-D conversion characteristics

Sampling frequency -	32 kHz
Initial resolution -	14 bits/sample
Companding -	near-instantaneous to 10 bits per sample over a 1 ms sample block
Coding for compressed samples -	2's complement + one parity bit
Channel bit rate -	352 kbits/second
Overall bit rate -	728 kbits/second

6.1.3 Data frames

In order for the data to be successfully decoded, it is necessary to provide some indication as to where the sound samples begin and end, and which belongs to which stereo channel. The data therefore is formed into *frames* which are made up as follows.

Each frame commences with a unique 8-bit *frame alignment word* — rather like the teletext framing code — which indicates the start of a new frame. The next 5 bits contain control information to indicate the form of the digital data.

The 5 bits have functions as follows:

C_0 is called the *frame flag bit* and is set to logic 1 for 8 successive frames and logic 0 for the next 8 frames, and thus indicates a 16-frame block. The function of this bit is to synchronise changes in the type of information being transmitted. C_1-C_3 are used to indicate the type of information being transmitted. At present four variations have been specified — one stereo audio channel, two independent monophonic channels, one monophonic channel and one data channel, and one data channel (704 kbits/second).

C_4 is called the *reserve sound switching flag* and is used to indicate that the receiver should switch to the conventional sound signal in the event of failure of the digital signal. This would be the case if the conventional sound signal is the same (or a monophonic version of) the digital sound signal.

Following the control bits are 11 data bits which are presently unspecified and have been reserved for future applications.

The data proper then follows as a block of 704 data bits — that is, one data block from each channel (the data for 1 ms of sound or 352 kbits of data for each channel). Thus the overall bit rate is $8 + 5 + 11 + 704 = 728$ bits/second.

The order of the data in this block depends on the type of data being transmitted; for a stereo signal the samples are interleaved, but for the other variations the sound/data channels are transmitted in alternate frames (two blocks per frame).

Figures 6.1 and 6.2 illustrate the composition of each data frame for each of the above data types (before interleaving — see below).

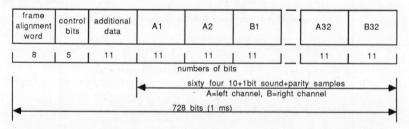

Figure 6.1   Data frame for a stereo sound signal (before interleaving)

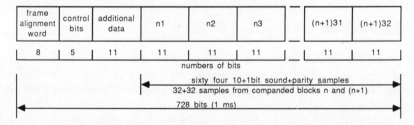

Figure 6.2   Data frame for two independent sound/data channels (before interleaving)

In order to minimise the effect of multiple errors, the bits are not transmitted in sample order, but are interleaved such that adjacent bits before interleaving are separated by at least 16 other bits after the interleaving process is completed. Again the details of this process are not important.

Finally before transmission, in order to ensure the signal energy in the modulated signal is spread evenly across the allocated band, after the frame alignment word all bits are scrambled by being exclusive OR-ed with the output of a pseudo-random sequence generator. This generator is duplicated in the receiver, and is initialised to a pre-defined state at the start of each frame. Thus the data can be de-scrambled successfully in the receiver.

The final resulting frames are then modulated for inclusion in the composite video signal as explained below. This method of coding, assembly and processing of the sound/data signals at first appears very complicated to implement (and comprehend!). However, using specially designed integrated circuits, the functions needed to perform the processing and subsequent recovery of the signals can be implemented quite easily and, importantly for the receiver manufacturer, cheaply.

6.1.4 Modulation method

Experiments have shown that by placing the second sound sub-carrier at a frequency just above the conventional sound carrier, a bandwidth of about 700 kHz can be made available. In order for a data rate of 728 kbits/second to be transmitted within this bandwidth, differentially encoded quadrature (4-phase) phase shift keying (QPSK) is used. With this system, the bit stream is divided into pairs, and each of the four possibilities for the states of the bit pairs causes a phase shift of the carrier by one of the four quadrature phases ($0°$, $+90°$, $-90°$ and $180°$). Using QPSK means that the bit rate of 728 kbits/s can be achieved within a 700 kHz bandwidth as well as providing unambiguous reception of the data, since the phase shift of the carrier for each bit pair is related to the current phase, not to an absolute phase reference.

The carrier frequency used is positioned just above the conventional sound carrier; the actual values chosen are 5.85 MHz for the European 625-line system and 6.552 MHz for the British 625-line system. A value for the 525-line system has not at present been specified. The carrier level is specified to be 20 dB below the peak white carrier level for the video signal.

Figure 6.3 shows how the additional digital sound carrier fits into the 8 MHz bands used for transmitting the video signals in the British 625-line system as explained in chapter 1. Notice that the new sub-carrier does actually overlap with the vestige of the upper adjacent channel. Tests have shown, however, that as long as the amplitude of this carrier is kept low enough, the interference caused is minimal — this is why the amplitude of the carrier is specified to be 20 dB lower than the maximum video carrier amplitude. Also transmitters are usually arranged such that adjacent channels occur on transmitters which are placed as far apart as possible, so that interference is in any case not generally a problem.

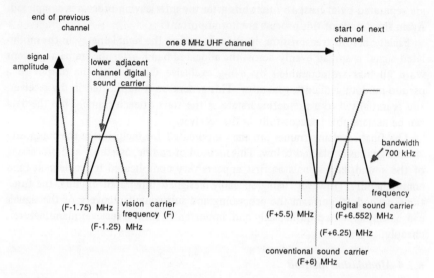

Figure 6.3 UHF channel characteristics with the new digital sound carrier

6.1.5 Summary

The specification for a digital stereo sound channel represents an exciting step forward in the transmission of television sound. Not only does it offer a high-quality stereo sound channel, but also the possibility of two independent sound channels for dual language broadcasting or a mixture of sound and data channels.

At the same time that this specification was being prepared, the MAC family of systems was also being developed, and thus the method of quantisation and processing of the digital sound data has been kept compatible as far as possible. This means that common integrated circuits can be developed for both systems.

At the time of writing, field trials of the system are in progress — notably in Britain and the Nordic countries — and the integrated circuit manufacturers are starting to produce decoding circuits for receivers.

6.2 The MAC system

The MAC system (Multiplexed Analogue Components) was designed primarily for the transmission of television pictures directly to peoples' homes via satellites — known as DBS (Direct Broadcasting by Satellite). Its use is now also being considered for distribution of video and data over cable networks.

One of the main considerations for the MAC system is that it should produce a picture quality equal to or better than that obtainable with conventional systems, and that it should be flexible.

At the time of writing (1988), public service DBS is not yet operational, and although it is certain that the MAC system will eventually be used, individual countries have not yet finalised on the precise variant they will use (see later).

6.2.1 The basic MAC system

The basic MAC system, as its name implies, involves transmitting each of the components of the television signal separately in different time slots.

The basic line format is retained (625 or 525-lines, 2:1 interlace and 4:3 aspect ratio), however each scan line is divided into three sections.

The first is the line blanking period which may, depending on the system variant, contain the sound information (in digital form) and other general data. The second section which occupies one-third of the active line period contains a time-compressed version of one of the colour difference signals, and the remaining two-thirds contains a time-compressed version of the luminance signal for that line. The two colour difference signals are transmitted alternately on consecutive lines, just as in the SECAM system.

The basic MAC signal, shown in figure 6.4, is designed to be frequency modulated on to a suitable carrier so that it occupies one 27 MHz satellite channel as internationally agreed in 1977. However, its use at baseband (unmodulated) or modulated on a lower carrier for distribution by cable is now being considered.

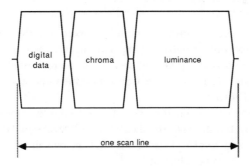

Figure 6.4 Basic MAC signal

The main advantage of this system is that the luminance and chrominance signals are completely separate, and thus there are no cross-colour or cross-luminance effects. Also, varying the compression time slots and/or ratios allows the possibility of transmitting pictures with a higher definition and/or a different aspect ratio.

There are four variants of the system which involve different methods of transmitting sound (and data) information with the video signal. These are designated A-, B-, C- and D-MAC and they cover the four possibilities of transmitting the sound/data at baseband or radio frequency (RF) using time or frequency multiplexing as shown in table 6.2.

Table 6.2 MAC sound/data format options

	Frequency multiplex	Time multiplex
Baseband	A	B
RF	D	C

6.2.2 The MAC variants

A-MAC

This variant provides frequency multiplexing of the sound/data (digital or analogue) at a baseband frequency just above the highest frequency component of the compressed video signal.

The advantage of this variant is a very rugged sound channel. However it has limits on the data rate and thus the number of sound channels, and on its potential for increasing the video bandwidth for extended definition as explained above.

B-MAC

The sound/data (in digital form) is time multiplexed into the section of the line period normally reserved for line blanking. This means that the receiver will need to contain storage so that the digital sound can be re-created as a continuous signal. However such techniques are well established and easy to achieve with modern integrated circuits.

To maximise the use of this time slot, the wide dynamic range of the 'video' signal is utilised by transmitting the data as a multilevel signal (that is, there are more than two logic levels). This means that an overall data rate of around 1.8 megabits/second can be achieved.

C-MAC

C-MAC again uses the line blanking period to contain the digital sound and data, however this variant uses direct phase shift keying of the high-frequency FM carrier.

The advantage of this method is that it allows a very fast data rate of above 20 megabits/second. However a baseband bandwidth of about 10 MHz is required which may make it unsuitable for cable applications.

D-MAC

The fourth variant uses frequency multiplexing, but at the FM carrier frequency rather than at the baseband frequency as in A-MAC. This variant allows the video and data to be separated, and high data rates are also possible. However separate receivers are needed, and interference between the video and data can become a problem.

As can be seen, each of these variants has advantages and disadvantages. In Britain, C-MAC was originally favoured, but D-MAC (as a reduced bandwidth sub-variant called D2-MAC) is now being seriously considered because of the revival of interest in cable systems and because of its probable adoption by France and West Germany for satellite broadcasting. In the USA and Australia, B-MAC is the favourite mainly because the baseband bandwidth in the 525-line system can be held to just over 6 MHz and can thus be used over the many cable systems in use in the USA. Whichever variant is eventually used, the European Community has directed that only MAC systems may be used on satellite broadcasting, encouraging further development of the MAC system.

6.2.3 Synchronisation

The MAC signal does not contain the conventional line and field synchronising pulses, although a line structure is still used. Synchronisation is achieved by digital means as follows. The start of each line contains a 6-bit data burst which has one of two values (one is the inverse of the other). These data bursts alternate on consecutive lines but each field ends with two consecutive identical bursts, the data being different for odd and even fields. The last line of each picture (line 625 in the 625-line system) is reserved for a frame synchronising signal. In this line, after the line sync burst, there is a set of 96 bits which is divided into a clock run-in period of 32 bits followed by a unique 64-bit frame synchronising word. Identification of the colour signal (red or blue difference) is achieved simply by stating that odd lines contain the U component, and even lines contain the V component.

Thus synchronisation is achieved totally by digital means which is clearly very different from the conventional line and field sync pulses designed to be decoded with analogue circuitry. Synchronisation by this means results in only 0.2 per cent of the total time being used for synchronisation rather than 20 per cent as in conventional systems.

6.2.4 *Control data*

Following the frame sync data on the last line of each frame is data that indicates the format of the MAC line and other information about the contents of the data signal. It is in this data that the flexibility of the system lies, since simply by changing the data, the system can be used to transmit higher definition pictures, or more sound channels (up to 8 separate sound channels can be carried by C-MAC) or a completely different combination of data and video.

The data in the last line contains:

(a) Information identifying the sound channel(s).
(b) Information describing the form of the time division multiplex (for example, television, high definition television, no television only sound/data etc.).
(c) Universal time for use by external equipment (video recorders etc.).
(d) The list of services available on the satellite channel (television, radio, teletext etc.).
(e) The description of the programmes being transmitted — including the language used.
(f) The local time in an ASCII character form.
(g) Data describing the form of each sound channel, that is
  • audio bandwidth
  • mono or stereo
  • linear or companded coding
  • level of error protection
  • music or speech

Figure 6.5 shows the composition of each of the lines in the MAC signal — note that the active parts of the spare lines in the field flyback interval are still available for teletext data.

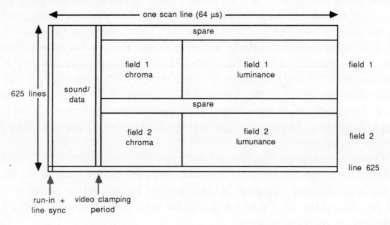

Figure 6.5 The MAC signal at frame rate

6.2.5 Time compression

One of the most difficult aspects of the system to comprehend at first is the idea of time compression, and the bandwidth of the resulting signal.

It was mentioned above that the luminance signal was compressed by one-third — that is, the luminance information in one active line scan period needs to be transmitted in a time equal to two-thirds of this active time.

This is achieved by sampling the video signal during the active line scan at time intervals of t_s seconds but transmitting these samples (still in analogue form) at intervals of $\frac{2}{3} \times t_s$ seconds. This directly increases the bandwidth of the resulting compressed signal by the ratio 3/2.

Thus for the 625-line system, the time-compressed luminance bandwidth will be increased from the uncompressed value of 5.5 MHz to $\frac{3}{2} \times 5.5 = 8.25$ MHz and for the 525-line system the luminance bandwidth will be increased from 4.2 MHz to 6.3 MHz.

The same operation is carried out on the chrominance signal; however the original bandwidth before compression is about one-half that of the luminance signal. Thus by time-compressing this signal by twice that of the luminance signal, the resulting signal occupies one-half the time of the compressed luminance component (the remaining third of the active line period), and has a bandwidth equal to that of the compressed luminance signal.

By the use of electronic memories (analogue or digital), the chrominance and luminance signal samples are read out to form the video section of the MAC signal. In the 625-line system, a 20.25 MHz clock is used to form the time-compressed signal; figure 6.6 shows the number of clock periods used in each of the sections of the MAC signal (for the C-MAC system where the data is also time multiplexed).

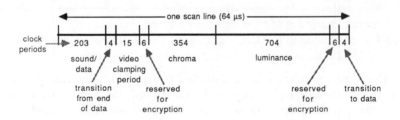

Figure 6.6 Composition of the C-MAC line

Note that there are periods of between 4 and 7 clock periods between the sections, and there is also a 15-period section to provide a reference level (clamp level) for the video signals.

6.2.6 Sound coding

Although with the A and D variants of the MAC system the sound could be transmitted as a straight analogue signal and with the B and C variants as a time-compressed analogue signal, it is most likely that a digital system will be employed for all variants. Digital sound technology is well established and the techniques for transmitting high-quality sound as a serial data stream are well known.

In order to allow for more than one sound/data channel to be transmitted simultaneously, a system of data packets is employed similar to the frames in the stereo sound transmission system. Each packet consists of a header which contains an address, a continuity index and a set of error-correction bits followed by the packet data proper.

The address defines the particular sound or data channel (television sound, radio, language, subtitle etc.) and the continuity index is a 2-bit counter that is clocked on each subsequent packet with the same address – this allows the receiver to detect lost or duplicated packets.

There are two types of sound channel that can be specified – high-quality (bandwidth 40–15 000 Hz) and low-quality (bandwidth 40–7000 Hz); their respective sampling frequencies are 32 kHz and 16 kHz. In addition, a high-quality channel can be monophonic or stereophonic. The samples are coded into 14 bits per sample, and there are four options on how the data is transmitted:

(a) Linear coding with simple parity protection. This results in a data rate of 480 kbits/s and allows up to six high-quality sound channels.
(b) Linear coding with Hamming error correction. This gives better performance under noisy conditions, but results in a data rate of 608 kbits/s or up to four high-quality sound channels.
(c) Near-instantaneous companding with simple parity protection. Using 10-bit samples from the 14 bits by companding – as in the stereo sound transmission system – the data rate is reduced to 352 kbits/s and provides the maximum number of eight high-quality sound channels.
(d) Companding as in (c), but with Hamming error protection, gives a data rate of 480 kbits/s or up to six channels of high-quality sound.

Using the control data in the last line, a mix of high/low quality channels using linear/companded coding can be transmitted, the only limitation being that the combined data rate must not exceed the maximum allowed for the MAC variant in use. General unspecified data can also be incorporated into the packets – as in the stereo sound system. The uses for the sound channels include stereophonic high-quality television and radio, low-quality commentaries and/or alternative language dubbing etc. There is also provision for mixing two of the channels so that music, sound effects etc. can be mixed with different language speech to enable television programmes to be truly international.

6.2.7 C-MAC data packets

As an example of the manner in which these packets are incorporated into the line multiplexed signal, the method used in the C-MAC system variant will be mentioned. In this variant, packet length is 751 bits of which 23 are used for the header. In each line multiplex, there are 198 data bits organised as two consecutive sets of 99 bits (this arrangement is to retain compatibility with the D2-MAC variant which halves the data rate and only contains one set on each line – thus halving the bandwidth required). This allows 162 packets of data to be transmitted in every frame, and this gives an overall bit rate of 3.04 megabits/second. This is illustrated in figures 6.7 and 6.8.

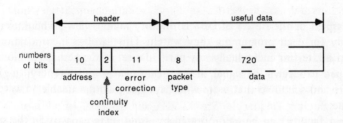

Figure 6.7   Data packet structure

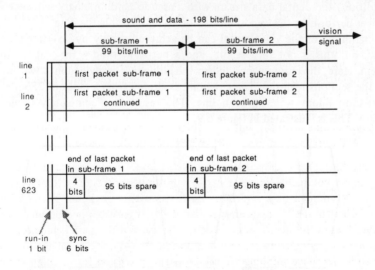

Figure 6.8   Sound/data transmission in the C-MAC variant

Packet synchronisation is achieved by defining the start of a new packet as the first bit following the line sync word on line 1. Packets then follow sequentially through the scan lines; two sets of 95 bits are left spare at the end of line 623 and the next packet starts at line 1 of the next scan frame (this means that packets are divided across line scans). Thus a frame alignment word is not needed.

Bit interleaving and scrambling is applied to each packet before transmission to protect against multiple errors and for energy dispersal respectively. The method adopted is very similar to that used in the stereo sound system.

6.2.8 Scrambling

Another aspect of the future of DBS is that of 'pay-per-view', where users pay individually for each programme they watch. This involves incorporation of a method of addressing individually any particular set of the perhaps millions that are equipped to receive the signal, and a means of scrambling (encrypting) both the picture and sound so that users who have not paid are unable to watch the programme.

The first facility can be incorporated by using spare capacity in the sound/data signal and, as long as the addressing system is made sufficiently large, then any number of users can be individually addressed.

Scrambling of the sound can easily be achieved by exclusive OR-ing (modulo 2 addition) the digital data stream with a pseudo-random binary sequence — the receiver performing the opposite function to de-scramble the data.

Scrambling of the video signal can be achieved simply by inverting the video signal, but this is not very secure, and a better method is the 'cut-and-rotate' system. Here, the chrominance and/or the luminance is divided into two sections and the sections reversed. The cut point is altered from a range of 256 points in a pseudo-random way on each line. The result is a completely unviewable picture. This is illustrated in figure 6.9.

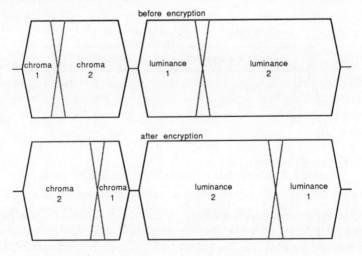

Figure 6.9 Video encryption by 'cut-and-rotate'

The cut operation produces extra transitions when re-assembled, so the sections are made to overlap and thus the transition can be gated out before display.

6.2.9 Summary

The MAC system is another exciting development in the field of television, and represents the most significant development in this field since the introduction of colour. The system uses the most up-to-date technology, but is designed such that MAC receivers can be built easily and cheaply using modern large scale integrated circuits.

As explained above, at present no MAC system is in commercial operation, but when the details of the specifications are finalised, and the integrated circuit manufacturers start to produce decoding devices, then DBS will become a reality and will open up a whole new range of truly international television, radio and data services.

Table 6.3 summarises the main features of the principal MAC variants which are being considered.

Table 6.3 Summary of the principal MAC variants

Variant	Modulation method	Modulated bandwidth MHz	Luminance bandwidth MHz	Data rate Mbits/s	Data format	Audio channels
C-MAC	FM	27	5.6	20	QPSK	8
B-MAC	FM	27	5.6	14	4-level	6
D-MAC	AM	12	5.6	20	3-level	8
D2-MAC	AM	10	5.6	10	2-level	4
D2-MAC (reduced bandwidth)	AM	7	4.5	10	2-level	4

6.3 The World System Teletext specification

The World System Teletext specification (WST) is a development of the original teletext specification described in chapter 3. Both compatibility and reverse compatibility have been retained, and, as far as possible, compatibility with the extended videotex specification described in the next section has also been retained.

6.3.1 The need for enhancements

Ever since the early days of teletext there have been comments about the limited graphic capability, the restrictive attribute system whereby attribute changes occupy a character space, and the lack of alternative character sets.

The need for a new teletext specification only became really necessary when these extended features were specified for videotex systems and were becoming available on videotex services. It was considered that since, from the start, the display system of both teletext and videotex were made compatible, this compatibility should be continued as far as possible.

Also, because of the 'information technology revolution', possible commercial applications for teletext carrying data services and telesoftware have recently been identified.

6.3.2 525-line systems

While extending the original teletext system, the opportunity was taken to modify the specification to include teletext on 525-line systems.

Because of the smaller video bandwidth in these systems, the bit rate needed to be reduced (to 5.727 272 megabits/second) and thus only 37 bytes of data are able to be transmitted on each teletext television line.

To enable displays of 40 character rows to be used, the 3-bit magazine number is reduced to 2 (allowing only 4 magazines), and the remaining bit used as a 'tabulation' bit. When this bit is set to zero, the packet contains the data for the first 32 characters in the row. When the tabulation bit is set to 1, the packet number is the first of a group of four display rows, and the data in the packet contains the last 8 characters in each of these four rows. Thus to specify a whole page of 24 rows requires an additional 6 teletext packets with a 525-line system.

6.3.3 Levels

The World System Teletext specification is divided up into five presentation levels. The features of each of these levels are summarised below. Each presentation level includes the features offered by all the lower levels.

Level 1

This is the original teletext specification described in chapter 3. In addition, level 1 provides for page links to be specified with each page, rather like videotex links, and also for broadcast service data to be transmitted. This data includes network identification, universal time, television programme details and a status message. It is independent of any other teletext data and is transmitted approximately once a second.

Level 2

In addition to the level 1 features, this level provides for:

(a) Up to 8 different options for the 'national option' positions in the basic character set.

(b) Up to 32 different colours — for each of these, each of the three basic colours (red, green and blue) is defined in one of 16 levels (giving a total of up to 4096 colour possibilities).
(c) Non-spacing attributes — that is, colours etc. can be changed individually on any character space without the need for an earlier control code space.
(d) Full row and full screen colours — that is, the areas outside the normal display area.
(e) Accented and supplementary characters.
(f) Smoothed mosaics.
(g) Double width and double size characters.
(h) Additional flash functions including 3-phase flash to give the impression of movement.
(i) Scrolling of additional teletext pages into a defined area of the currently displayed page.
(j) Dynamic redefinition of the thirteen 'national option' positions in the basic character set.
(k) Dynamic redefinition of sixteen of the 32 possible colours.
(l) Redesignation of the basic character sets to provide the possibility of a whole range of alternative characters (codes have been reserved for Latin, Arabic, Cyrillic, Devanagari, Greek, Hangul, Hebrew, Burmese, Lao, Malayalam, Singhalese, Thai, Katakana, Hiragana, Chinese, Tamil, Assamese, Gujerati, Punjabi, and Telegu character sets).

Figure 6.10 is a photograph of a sample level 2 teletext page showing some of the features mentioned above.

Level 3

This level additionally defines a method of implementing dynamically redefinable character sets (DRCS) which enables any character or graphic shape to be displayed within the limits of the pixel size.

Figure 6.11 is a photograph of a sample level 3 teletext page where the motifs are built up using DRCS characters.

Level 4

This level is reserved for alphageometric displays (like the geometric displays described in the CEPT videotex section) which are currently being standardised.

Level 5

This level is reserved for alphaphotographic displays (like the photographic displays described in the CEPT videotex section) which are currently being standardised.

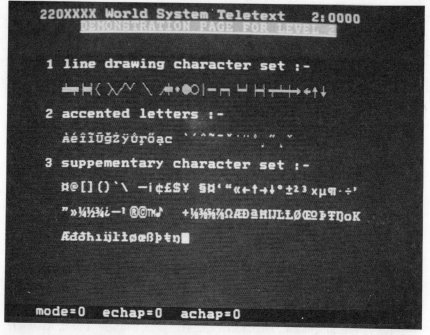

Figure 6.10 Sample level 2 teletext page

Figure 6.11 Sample level 3 teletext page

In addition to the above levels, the specification contains details on:

(a) A method of transmission of data for processing — which includes data for further processing outside the teletext decoder (telesoftware) as well as level 4 and 5 data.
(b) A method of conditional access to a teletext service — pages can be scrambled before transmission, and a 'page key' is used to decipher the information. This allows access to certain pages to be limited to certain users.
(c) A method of implementing independent data services. These are data services which are not connected and do not interfere with the main teletext information.

6.3.4 Implementation

One of the most important provisions of a technical specification such as that for the teletext service is that it should not define every detail so completely that there is no room for future expansion. The original specification allowed for such expansion in many places, and this has enabled the World System Teletext specification to be produced. Another important provision is that any enhancements should not affect the operation of existing systems and such compatibility has been maintained in the new specification.

It is a credit to the authors of the new specification that this itself has provisions for future expansion — 'no action' or 'reserved for future expansion' occurring at many places in the specification.

The main 24 rows of display data are transmitted on packets 0 to 23, which requires 5 bits to specify the actual packet number. These 5 bits can, however, specify up to 32 different packets. These extra packets (packets 24 to 31) are ignored by teletext decoders operating to the original specification.

The new specification uses these spare packets to implement a great many of the new facilities mentioned earlier. This is how they are used:

Packets 24 and 25

These two packets are reserved for extra display rows. Packet 24 could, for example, be used to display information on an extra 25th row, and packet 25 could be used to overwrite the header row (packet 0). Alternatively, packets 1 to 25 and the display data in packet 0 provide a 1K block of data which could be used for the transmission of computer programs (*telesoftware*).

Packet 26

This packet is used to provide parallel attributes and/or to overwrite characters in the main display. After the magazine and packet address (row number), this packet contains a 'designation code' which allows up to 16 different packet 26s to be transmitted. The remainder of the packet contains 3-byte 'triples' which contain an address, mode and data to specify the required parallel attribute or overwrite character and its position in the displayed page. Figure 6.12 shows the format of packet 26.

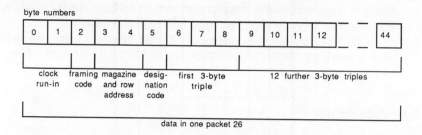

Figure 6.12   Format of packet 26

The functions available using data in packet 26s are:

(a) Set full row and screen colours.
(b) Set cursor position.
(c) Define scrolling area.
(d) Set foreground and background colours.
(e) Overwrite with a character from a supplementary character set (additional symbols (figure 6.13), smooth mosaics and lines (figure 6.14). DRCS character etc.).
(f) Additional flash controls (different flash rates, 3-phase flash which creates the impression of movement, inverted flash).
(g) Set parallel attribute (double height/width/size, conceal etc.).
(h) Set latching shift to a different character set.
(i) Overwrite with an accented character.

Packet 27

This packet is used to provide page 'links'. Like packet 26, this packet contains a designation code and two sets of packet 27s have been defined:

(a) *Designation codes 0 to 3.* These provide up to 24 page links, rather like the links on a videotex page. The teletext editor can thus supply links to other relevant pages in the teletext service and so make the service more 'user friendly'. Special user keys can be supplied to invoke acquisition of the linked pages, or alternatively some or all of these pages can be automatically acquired and stored in the teletext decoder to be instantly available when the user has finished reading the currently displayed page.

	0	1	2	3	4"	5	6	7
0				°		—	Ω	ĸ
1			¡	±	`	¹	Æ	æ
2			¢	²	´	®	Đ	đ
3			£	³	^	©	ª	ð
4			$	×	~	™	Ħ	ħ
5			¥	µ	¯	♪		ı
6			#	¶	˘	₠	Ĳ	ĳ
7			§	·	˙	‰	Ŀ	ŀ
8			¤	÷	¨	∝	Ł	ł
9			‘	’			Ø	ø
10			“	”	˚		Œ	œ
11			«	»	¸		º	ß
12			←	¼	_	⅛	Þ	þ
13			↑	½	˝	⅜	Ŧ	ŧ
14			→	¾	˛	⅝	Ŋ	ŋ
15			↓	¿	ˇ	⅞	ŉ	

Figure 6.13 Latin supplementary character set

(b) *Designation codes 4 to 7*. These packets again provide up to 24 page links, but the pages that are specified are not necessarily pages for direct display. Such pages are called 'pseudo-pages' and the packet 27 links associated with the currently displayed page can indicate the following types of linked page.

(i) not a pseudo-page (that is, a normal page for display);

(ii) a pseudo-page which is completely to overwrite the currently displayed page;

(iii) a pseudo-page containing information to be scrolled into the currently displayed page;

(iv) a pseudo-page containing the specification for a set of DRCS characters;

(v) a pseudo-page which can extend both horizontally or vertically the size of a page (the maximum size can be 160 columns and 101 rows per page – such a page would of course require a number of pseudo-pages to be completely specified);

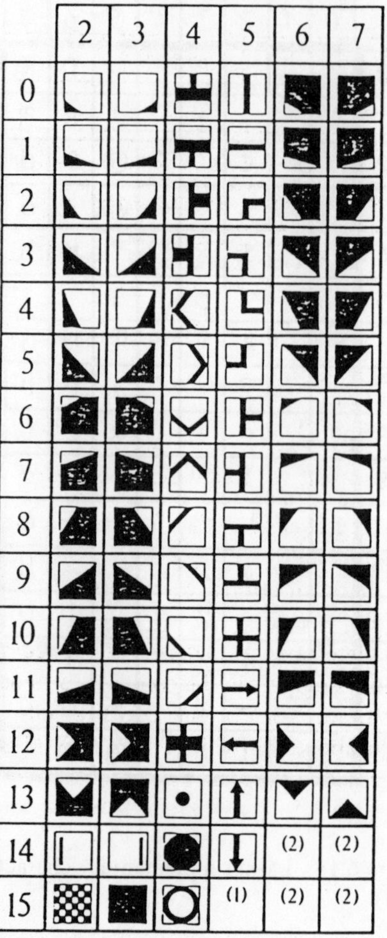

(1) This character permits the display of background colour
(2) No character assigned

Figure 6.14 Smooth moasic and line character set

(vi) a pseudo-page which requires reformatting using some agreed protocol between the originator and receiver of the data (such as scrambled pages for a conditional access teletext service as mentioned earlier).

Packet 27s with designation codes 8 to 15 have not been defined.

Both types of packet 27 can, in addition, indicate that the page is at the start, middle or end of a chain of associated pages. Figure 6.15 shows the format of a packet 27.

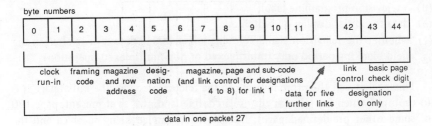

Figure 6.15   Format of packet 27

Packet 28

This packet is used to provide the following functions:

(a) It indicates the type of page — that is, a normal page or a pseudo-page of one of the types given above.
(b) It allows the 13 'national option' characters in the primary character set to be redefined.
(c) It allows the complete basic character sets to be redesignated.
(d) It allows 16 of the 32 colours in the colour map to be redefined.
(e) It allows the four locations of the DRCS colour pointer table to be redefined (a four location table which enables DRCS characters to use four of the 32 colours defined in the main colour table).

Packet 29

This has the same form as packet 28, but defines the code tables, colours etc. for a particular magazine, rather than an individual page.

Packet 30

This is presently left undefined except for one special case, that is packet 30 of a page in magazine 8 (known as packet 8/30). This packet is the 'Broadcast Service Packet' and is independent of any other page data. It is specified to be transmitted approximately once a second (being a single packet, it only occupies one television scan line) and contains the following information:

(a) a page link specifying the initial page that the teletext decoder is to acquire on switch on;
(b) a network identifier;
(c) the current time and date;

(d) a television programme label;
(e) data for initial display on switch on;
(f) whether the television signal contains full channel teletext or teletext in the field blanking period only (multiplexed or non-multiplexed operation).

This packet can therefore indicate to the teletext decoder the initial page it is to capture (which the teletext editor can define, and not, as at present, page 100 or some other pre-defined page), the television service being received and the television programme being shown. This information could be used to, say, automatically tune a receiver to a particular channel, or switch on and off a video recorder to record a particular programme.

Packet 31

This packet is used to implement independent data services. Recently, the potential for using teletext as a means for distributing data nationally and at speed has been realised. However all the packets except for packet 8/30 are related to a particular page, and thus cannot be used to carry data which is not associated with a specific page. It would, of course, be possible at the transmitting end to format the data into pages, but for some data, this may not be convenient or practical. Packet 31 however is defined to be page independent and can be transmitted at any time without affecting the correct operation of the basic teletext service.

This packet contains the following information:

(a) up to 15 different data channels;
(b) various different optional data formats and indicators:
repeat facility
include a continuity indicator
include a data length indicator
(c) service packet address – allowing individual users or groups of users within a channel to be identified;
(d) user data – between 28 and 36 data bytes (20 and 28 for 525-line systems) depending on the amount of optional initial data above;
(e) cyclic redundancy check word.

As can be seen, these extra packets have been used to implement a wide range of extra facilities. The details of how each packet is actually constructed is very complex (many of the packets contain Hamming encoded data in various formats and/or cyclic redundancy check words to reduce the chances of error for important data etc.). It is hoped, however, that the above description will give an overall idea on how these packets are used to provide the additional facilities.

6.3.5 Progress on implementation

Recently IC manufacturers have started to produce display devices capable of producing World System Teletext type displays, and there are some laboratory prototypes of teletext decoders which incorporate most of its features (most of levels 1, 2 and 3). The problem here is not only in the design of a commercially viable decoder, but in the original creation of the teletext pages at the transmission end.

To create such pages requires sophisticated editing equipment and equipment to convert the pages into data conforming to the specification, as well as the personnel to design the pages and operate the equipment.

At present, in Britain, the two broadcasting authorities do not have the resources to invest in such a venture. A similar problem exists with television manufacturers and thus there is a 'chicken and egg' situation.

In Britain there has, however, been an attempt to implement commercially some of the WST features and this basically involves keeping the display system unchanged (that is, only serial, spacing, attributes and limited character sets) but adding the packet 27/0-27/3 links. This has become known as *Fastext*.

The idea in Fastext is to supply an additional set of four 'soft' keys on the user's keypad and to use the extra 25th display row, via packet 24, to give, for each page, a 'menu' of associated pages which can be accessed directly by these keys. An additional 'index' key is also provided to allow direct access to the current relevant index page. These five page links are contained in one packet 27 and can be pre-captured ready for instant display.

Figure 6.16 is a photograph of a teletext page which shows the extra 25th display row giving a menu of options available with the four soft keys. The menu is colour coded to correspond with the colours of the four keys as well as giving the options in the same order as the layout of the keys on the user's keypad. Identification of the keys in the menu is thus implied and does not have to be stated explicitly.

Also in Fastext is the inclusion of the Broadcast Service Packet which allows the teletext editor to have control over which page should be captured when the decoder is switched on as well as the other features detailed in the packet 8/30 description above.

The main advantages of using Fastext as a stepping point to a full implementation of the specification are:

(a) The broadcasters need only supply minimal extra resources and can use existing equipment.
(b) Current second generation teletext decoder ICs can be used.
(c) Only a minimal increase in page storage capacity is required in the television set.
(d) User friendliness is increased.
(e) Apparent access time is dramatically reduced (even with a four page store).

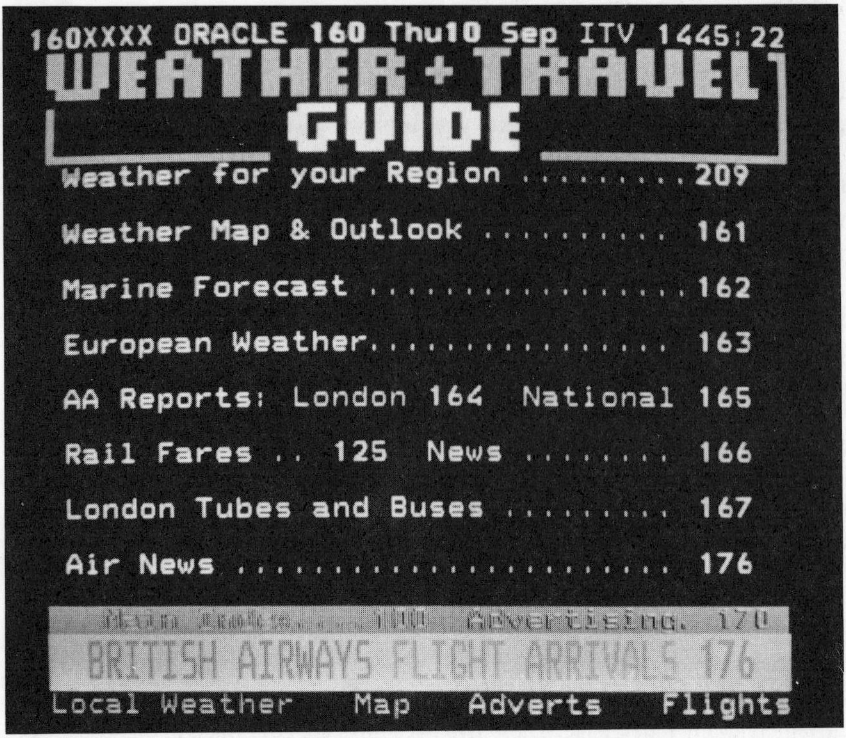

Figure 6.16 Sample Fastext teletext page

6.3.6 Summary

The WST is a major step forward in the teletext system. It opens up the possibility for a truly international specification which is capable of transmitting text in nearly any language. The graphics capability is also enhanced, and in the near future the specification will be extended to include sophisticated line drawings and pictures.

Although at present no country has fully implemented the new specification, it will not be long before new dedicated decoder integrated circuits are available, and the broadcasters and information providers will start to provide data to the WST standard.

The use of teletext as a fast and wide data distribution system also has exciting possibilities, and this aspect is presently causing much interest. In Britain, the BBC and IBA have already set up independent data services – the BBC (called *Datacast*) uses packet 31, and the IBA uses a page-based system.

6.4 The CEPT Videotex specification

As mentioned earlier, enhanced videotex systems were developed before the teletext system. The original videotex system was upgraded in the early 1980s to provide a common international specification that would encompass all the systems in operation then. These included the British Prestel system, the Canadian Teledon system, the German Bildschirmtext service, the French Teletel system, the Japanese Captain system and the British 'Picture Prestel' experimental system (alphaphotographic displays).

The Conference of European Posts and Telecommunications (CEPT) looked at all these systems which had many similarities in their operation, and produced a common specification – although the specifications for the alphageometric displays (pictures made up of lines, arcs and circles etc.) and alphaphotographic displays (pictures made up from specifying luminance and chrominance data for each pixel) have not yet been finalised.

The main additions to the basic videotex system described in chapter 4 lie in the enhanced display facilities. The basic structure of the database and the transmission system remain the same, except that the CEPT specification does allow for the transmission of data using 8-bit as well as the usual 7-bit format.

6.4.1 Presentation Protocol Data Units

The means of obtaining compatibility between the various systems described above lies in the use of Presentation Protocol Data Units (PPDUs). These are sequences introduced by control codes which contain two parts as shown in figure 6.17.

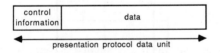

Figure 6.17   Presentation Protocol Data Unit

The first part of the PPDU contains the identification of the particular service required – alphamosaic (default), geometric (as in Teledon systems), photographic (Picture Prestel), sound, telesoftware and facsimile. The second part of the PPDU contains the data to be processed under the required PPDU.

The alphamosaic PPDU

This is the default PPDU and the terminal assumes this mode unless instructed otherwise. Under this PPDU the new specification provides for the display of different character sets, parallel attributes, DRCS characters etc., as mentioned in the previous section on the World System Teletext specification. Indeed the display systems are identical in most respects.

Additional control codes from columns 0 and 1 of the ASCII code table are used to invoke the following functions:
- Shift in (0/15): sets a locking shift to a different character set
- Shift out (0/14): restores the previously invoked character set
- Repeat (1/2): the previous character is repeated the number of times specified in the following bytes
- Cancel (1/8): all characters from and including the current cursor position to the end of the current line are deleted; the cursor is not moved
- Single shift 2 (1/9): the following character is taken from the currently invoked G2 character set (usually the Latin supplementary set)
- Single shift 3 (1/13): the following character is taken from the currently invoked G3 character set (usually the graphics characters)
- Active Postion Address (1/15): following data bytes define a new cursor row and column position

Figure 6.18 shows these new control codes together with the remainder of the cursor controls that were described in chapter 4.

row \ column	0	1
0	Nul	
1		Cursor on
2		Repeat
3		
4		Cursor off
5	ENQ	
6		
7		
8	AP back	Cancel
9	AP forward	Single shift 2
10	AP down	
11	AP up	ESC
12	Clear screen	
13	AP return	Single shift 3
14	Shift out	AP home
15	Shift in	AP address

AP=active position (cursor)

Figure 6.18 CEPT videotex control codes

Escape sequences (data bytes introduced by the Escape character (1/11)) are used to invoke the other additional display features as listed below:

- Different character sets (new character sets may be invoked and used just as if they were one of the default sets)
- Parallel attributes (these are the same as described in the WST specification above – the normal serial attributes are still available as well)
- Full row and screen colours
- Variable page format (different numbers of rows and/or columns)
- Page scrolling
- Redefining 16 of the 32 positions in the colour table from a 'palette' of 4092 possible colours (as in the WST)

The DRCS PPDU

This PPDU is used to define the shape (and colour) of special alphanumeric or graphic characters which, once stored in the terminal, can be invoked and used just as any other standard character.

The geometric PPDU

This PPDU is used to create line drawings which are composed of combinations of standard shapes such as lines, arcs, circles etc.

At present the format of this PPDU has not been finalised, but it is hoped that when specified, it will be compatible with that presently used in the Canadian Teledon system.

The photographic PPDU

Like the geometric PPDU, this has not yet been specified, but it will be used to implement video pictures (still television pictures) by specifying each pixel within the picture by its luminance and chrominance component.

In Britain this system has already been demonstrated by British Telecom with their 'Picture Prestel' experimental service.

The sound PPDU

This PPDU has been reserved to enable terminals to control sound generators by videotex codes.

The telesoftware PPDU

Use of this PPDU indicates that the data contained is not for direct display, and is to be used for processing by an external device. The data may be a computer program in machine code, or a low or high level language, or it may be data to be used by an external device.

The facsimile PPDU

This PPDU has been reserved to enable facsimile-like transmissions to be incorporated. This is to enable features of the Japanese Captain system (Character and Pattern Telephone Access Information Network) to be used in the future.

The Captain system is different because of the vast number of different character shapes available in the Japanese language. In the Captain terminal, a large character ROM is used to hold the 3000 or so most commonly used symbols, and the remainder are transmitted as direct character images in a facsimile-like manner.

At present no details on this PPDU have been specified, but it is hoped that a specification will emerge which will allow the CEPT specification to include the main features of the Captain system.

6.4.2 Overview

The CEPT videotex specification, although by no means complete, is an attempt to bring together the features of the various systems in use around the world into a common international specification.

Most of the features contained in the alphamosaic PPDU explained above are operational, notably in Germany with their Bildschirmtext service. Other European countries will soon be following and extending their services to include the new features offered by the CEPT videotex specification.

Like all the new specifications in this chapter, advances in technology have paved the way for the inclusion of new features and facilities.

6.5 Summary

This chapter has looked at some of the new specifications which, at the time of writing, are still in the process of finalisation, but will most certainly have a radical effect on television and teletext in the coming years.

As explained at the start, all these specifications are very complex and assume a detailed knowledge of communication techniques. It has only been possible therefore to give a very brief outline of their contents in this book. It is hoped though that such an outline will enable the reader to see the main features offered, and to understand in broad terms the method of implementation of these new features and their advantages.

Further Reading

BBC, IBA, BREMA (1974). *Specification of standards for information transmission by digitally coded signals in the field-blanking interval of 625-line television systems*, published jointly by the British Broadcasting Corporation, the Independent Broadcasting Authority and the British Radio Equipment Manufacturers' Association, London (October).

British Post Office (1980). *Prestel Terminal Specification*, 1st edn, British Telecom, London.

BVTTC (1986). *Code of Practice of Full Level One Features*, British Videotex and Teletext Technical Committee.

Carnt, P. S. and Townsend, G. B. (1961). *Colour Television, Volume 1*, Iliffe Books, London.

Carnt, P. S. and Townsend, G. B. (1969). *Colour Television, Volume 2*, Iliffe Books, London.

CENELEC (1981). *Domestic or similar electronic equipment interconnection requirements: peritelevision connector*, European Committee for Electrotechnical Standardisation (CENELEC) 2, rue Bréderode, Boîte 5, 1000 Bruxelles.

CEPT (1983). *CEPT Videotex Specification*, Conference of European Posts and Telecommunications recommendation T/CD 6-1.

Childs, G. H. L. (1983). 'The European Videotex Standard'. *British Telecom Technical Journal*, **1** (1), July.

Deutsche Bundespost (1983). *Functional Specification for Bildschirmtext Terminals*, Deutsche Bundespost, Fernmeldetechnisches Zentralanst, Section T25.

DTI (1986). *World System Teletext Specification*, Information Technology Division, Department of Trade and Industry, 29 Bressenden Place, London, SW1.

EBU (1984). *Television standards for the broadcasting satellite service – specification of the C-MAC/packet system*, European Broadcasting Union Technical Centre, Av. Albert Lancaster, 32, B-1180 Bruxelles.

EBU (1987). *Specification for transmission of two-channel digital sound with terrestrial television systems B, G and I*, European Broadcasting Union Technical Centre, Av. Albert Lancaster, 32, B-1180 Bruxelles.

Independent Broadcasting Authority (1983). *Developments in Teletext*, Technical Review No. 20, Independent Broadcasting Authority, Winchester, Hants.

Independent Broadcasting Authority (1984). *Light and Colour Principles*, Technical Review No. 22, Independent Broadcasting Authority, Winchester, Hants.

Lothian, J. S. and O'Neill, H. J. (1986). 'The C-MAC/packet system for satellite broadcasting', *Proceedings of the Institution of Electrical Engineers*, **133**, Pt. F, No. 4, July.

Lowry, J. D. (1984). 'B-MAC an optimum format for satellite television transmission', *Proceedings of the 18th Annual SMPTE Television Conference*, Montreal, February.

Mertens, H. and Wood, D. (1983). 'The C-MAC/packet system for direct satellite television', *EBU Technical Review*, No. 200, August.

Money, S. A. (1979). *Teletext and Viewdata*, Newnes Technical Books, Butterworth, London.

Mothersole, P. L. (1986). 'Developments in broadcasting technology and their effect on TV receiver design', *Electronics and Power*, March.

Pim, D. N. (1986). 'What's happening to Teletext?', *Electronics and Power*, February.

Reed, C. R. G. (1969). *Principles of Colour Television Systems*, Sir Isaac Pitman, London.

Sewter, J. B. and Wood, D. (1985). 'The evolution of the vision system for the EBU DBS standard', *Journal of the Institution of Electronic and Radio Engineers*, **65**(9), September.

Sims, H. V. (1969). *Principles of PAL Colour Television*, Newnes-Butterworths, London.

Smol, G., Hamer, M. P. R. and Hills, M. T. (1976). *Telecommunications: A Systems Approach*, George Allen & Unwin, London.

Index

Access time – teletext 60, 137
Active field factor 19
Active line factor 20
Active position 83
Acuity 3, 57
Adaptive data slicer 103
Addressing – page 61
Addressing data – teletext 68
AFC signal *see* Automatic frequency control signal
AGC *see* Automatic gain control
Alphageometric display 129, 139
Alphamosaic PPDU 139, 142
Alphaphotographic display 129, 139
A-MAC 120
Amplifier – scan signal 96
ASCII codes 63
Aspect ratio 19
Asynchronous data transmission 83
Attribute 61
   non-spacing (parallel) 129, 132, 140, 141
   serial 63
   spacing 63
Audio signal processing 92
Automatic frequency control (AFC) signal 89
Automatic gain control (AGC) 92

Back porch 17
Bandwidth 29
   chrominance 51
   chrominance (NTSC) 33
   chrominance, PAL 44
   composite video signal 19
   maximisation of 41
   PAL colour difference signals 44
   requirements for colour television 29
   telephone lines 86
   teletext signal 76

U and V signals 44
   video signal 11, 28
Bell curve 54
Bildschirmtext 139, 142
Bit interleaving 126
Bit rate – teletext 76
Bit synchronisation 67
Black level 17
Blanking level 17
Blanking signals 95
Blast-through alphanumerics 64
Block graphics 62, 66
B-MAC 120
Brightness 29
Brightness signal (Y) 30
Broad pulses – field 17
Broadcast service packet 135, 137
Bulk update – videotex 80
Burst 42
   MAC data 121
   sub-carrier 37, 41
   swinging 49, 50
Bus – inter-IC 110
   television control 110
Bus receivers 110
Byte synchronisation 67

Camera – colour 28
Camera tube 5
Captain 139, 142
Carrier frequency – digital sound 117
Ceefax 58
CEPT videotex 139
Character rectangle 61
Character ROM 104
Character set
   Latin supplementary 133, 140
   line 134
   smooth mosaic 134
   teletext 62

Chroma signal 33, 97, 98, 101
Chrominance bandwidth 51
  PAL 44
Chrominance signal 31, 51, 123
  NTSC 32
  separation in PAL 47
Clock run-in
  MAC 121
  teletext 103
C-MAC 120
Code — designation 131, 132
Code table
  teletext 63
  videotex 83
Colour 1
  cross 36
  monochromatic 2
  non-spectral 2
  pigments 26
  polychromatic 2
  primary 2
  spectral 2
  teletext 62
Colour camera 28
Colour decoder
  NTSC 98
  PAL 99
  SECAM 100
Colour difference signal 31, 51
  weighted 33, 40, 52, 56
Colour encoder
  NTSC 42
  PAL 50
  SECAM 57
Colour sub-carrier 33, 41
Colour television tube 28
Column 61
Companding 124
  sound 115
Compatibility 28, 57, 114, 127, 131
  effects in SECAM 56
  reverse 28
Composite video signal 15
  bandwidth 19
Contrast control 97
Control bits — teletext header 68, 72
Control bus — television 110
Control codes 62
  videotex 82
Control data — MAC 121
Control unit — television 107
Controller — videotex 106

Correction — gamma 14
Cross colour 36
Cut-and-rotate encryption 126

Data frames 115
Data packets — in the MAC system
  124
Data rate — teletext 77
Data services — independent 131
Data slicer — teletext 103
Data transmission
  asynchronous 83
  full duplex 80, 85
  half duplex 86
  serial 80
Datacast 78, 138
D_B signal 52, 101
DBS 118
Decoder
  NTSC 98
  PAL 99
  SECAM 100
  teletext 101
  videotex 104
De-emphasis
  SECAM 101
  sound signal 92
Delay
  I signal 41
  luminance 42
Delay line PAL 46
Designation code 131, 132
Difference signals — colour 31
Differential phase errors 43
Diode — varicap 89
Display
  alphageometric 129, 139
  alphaphotographic 129, 139
  characters 62
Display data — teletext 71
Display memory
  teletext 103, 105
  videotex 107
Display unit — teletext 103
D-MAC 121
Dot pattern 39
Double sideband modulation 34, 41
D_R signal 52, 101
DRCS *see* Dynamically redefinable
  character sets
DRCS PPDU 141
Dynamic range — of sounds 115

Dynamic redefinition 129
Dynamically redefinable character sets (DRCS) 129, 140

Electromagnetic waves 1
Encoder
  NTSC 42
  PAL 50
  SECAM 57
Encryption 126
Equalising pulses 17
Error correction 124
Escape sequences — videotex 83, 141

Facsimile PPDU 142
Fastext 78, 137
Field broad pulses 17
Field factor — active 19
Field flyback 5
Field frequency 12
Field scan 5, 24
  oscillator 96
Field sync pulses 15, 24
Filter — notch 36
Filtering of the UHF or VHF signal 88
Flag
  frame 116
  reserve sound 116
Flicker effect 10
Flyback
  field 5
  line 5
Flywheel synchronisation 95
FM 51, 52, 56
Frame 115
  alignment word 115
  flag 116
  synchronising signal — MAC 121
Framing code 68
  detector 103
Frequency
  choice of sub-carrier 38
  field 12
  picture 12
Frequency modulation 51, 52, 56
Frequency shift keying 85
Frequency synthesis 109
  tuning 90
Front porch 17
FSK *see* Frequency shift keying
Full channel teletext 78, 136

Full duplex data transmission 80, 85

Gamma
  camera 14
  system 14
  transmitted 14
  tube 14
Gamma correction 14, 24, 31, 40
Gamma correction circuit 14
Geometric PPDU 141
Graphics — block 62, 66
Guard band 22

Half duplex data transmission 86
Hamming error
  correction 124
  protection 69
Hamming parity checks 69
Hanover bars 46
Header 124, 125
  row 69, 71
Horizontal resolution 19
Hue 29
  changes 43
  control 43, 99
  errors 45

I signal 99
  delay 41
  NTSC 34, 41
I^2C bus *see* Inter-IC bus
Ident signal — SECAM 53, 55, 56
IF signal *see* Intermediate frequency signal
Independent data services 136
Information providers 80
Interference 59
Interference effects
  luminance/chrominance 35, 41
  SECAM system 56
Inter-IC bus 110
Interlace 15
Interlaced scanning 12, 24
Interleaving 117
Intermediate frequency (IF) signal 89
  processing 91

Kell factor 9, 19

Latin supplementary character set 133, 140
Level-dependent phase errors 43

Levels — teletext 127
Light 1
Line
  character set 134
  flyback 5
Line factor — active 20
Line scan 5, 24
  oscillator 95
Line sync pulses 15, 24
Linear coding 124
Lines
  number of 7
  unused 59, 60
Link — page 132
Links — teletext 128
Logic levels — teletext 74
Luminance
  delay 42
  signal 31, 40, 123

MAC
  A variant 120
  B variant 120
  C variant 120
  clock periods 123
  clock run-in 121
  control data 121
  D variant 121
  frame synchronising signal 121
  signal 119
  sound coding 124
  system 118
  time compression 122
Magazine 61, 73
Modem 79, 107
Modulation
  double sideband 34, 50
  frequency 51, 52, 56
  of the videotex signal 85
  quadrature 33, 40
  vestigial sideband 22, 34
Monochromatic colour 2
Monochrome 1

Noise 17
Noise performance 25, 36, 38
Non-spacing attribute 129, 132, 140, 141
Non-spectral colour 2
Notch filter 36, 55, 97
NRZ coding 67

NTSC
  chrominance signals 32
  composite video signal 31
  effects of sub-carrier phase errors 42
  signal parameters 35
  sub-carrier frequencies 40
  transmitter 42
Number of lines 7

Oracle 58
Oscillator
  field scan 96
  line scan 95
  sub-carrier 37

Packet 25, 131, 132, 135, 136
  broadcast service 135, 137
  data 124
  synchronisation 125
  teletext 66
Page 60
  pseudo 133, 135
  videotex 81
Page addressing 61
Page key — teletext 131
Page link 132
  teletext 128
Page stores — teletext 78
PAL
  choice of sub-carrier frequency 49, 50
  delay line 46
  simple 45
  system 42
  transmitter 50
Parabola generator 96
Parallel attribute 129, 132, 140, 141
Parameters — NTSC signal 35
Parity bit 67
  videotex 83
Parity checks — Hamming 69
Parity error 67, 104
Pay-per-view 126
Peak white level 17
Periconnector 111
Persistence — of the eye 11
Phase errors
  differential 43
  level dependent 43
  sub-carrier effects in the NTSC system 42

Index

Phase locked loop 90
Phase reversal of V signal in PAL 44, 50
Phase shift keying – quadrature (QPSK) 117
Phases of colour difference signals 33
Phasor
   colour difference signals 33
   I and Q signals 34
   NTSC burst 37
   PAL chrominance signals 45
Photocell 3, 4
Photographic PPDU 141
Photopic response 1
Picture frequency 12, 24
Picture Prestel 139, 141
Pigments – colour 26
Pixel 129, 141
Polychromatic colour 2
Porch
   back 17
   front 17
Power supply – switched mode 110
   television 110
PPDU (Presentation Protocol Data Unit) 139
   alphamosaic 139, 142
   DRCS 141
   facsimile 142
   geometric 141
   photographic 141
   sound 141
   telesoftware 141
Pre-emphasis
   FM sound 21
   SECAM sub-carrier 54
   video, SECAM 52, 56
Presentation levels – teletext 127
Presentation protocol data unit *see* PPDU
Prestel 79, 139
Primary colour 2
Pseudo page 133, 135
Pulse
   equalising 17
   field broad 17
   sandcastle 95, 98
   synchronising 15, 24

Q signal 99
   NTSC 34, 41

QPSK *see* Quadrature phase shift keying
Quadrature modulation 33, 34, 40, 50
Quadrature phase shift keying (QPSK) 117
Quantisation 115

Raised cosine waveform 76
Raster 6, 24
Receiver
   bus 110
   remote control 109
Redefinition – dynamic 129, 135, 141
Redesignation of character sets 129, 135
Remote control receiver 109
Remote control transmitter 109
Remote programming – of a videotex decoder 83
Reserve sound switching flag 116
Resolution
   horizontal 19
   of a television 61
   of photocell arrays 4
   vertical 19
Reverse compatibility 28
Rolling header 103
   display 72
Row 61

Sandcastle pulse 95, 98
Saturation 29
SAW filter *see* Surface acoustic wave filter
Sawtooth scanning waveform 93
Scan
   field 5, 24
   line 5, 24
Scanning 5
   interlaced 12
   rate 10
Scrambling 117, 126, 131, 134
SECAM
   compatibility effects 56
   encoder 57
   ident lines 53, 55, 56
   interference effects 56
   sub-carrier deviations 53
   sub-carrier phase 54
   sub-carrier pre-emphasis 54

SECAM *(cont'd)*
  sub-carriers 53
  system 50
  transmitter 57
Serial
  attribute 63
  data transmission 80
Service – teletext 60
Shadow mask 29
Signal
  automatic frequency control (AFC) 89
  chroma 33, 97, 98, 101
  chrominance 31, 51, 123
  D_B 52, 101
  D_R 52, 101
  I 99
  intermediate frequency (IF) 89
  luminance 31, 40, 123
  MAC 119
  MAC frame synchronising 121
  Q 99
  U 33, 100
  V 33, 100
  vertical insertion 60
  video 3
  videotex 83
  weighted colour difference 33, 40
Simple PAL 45
Smooth mosaic character set 134
Sound
  analogue-to-digital conversion 114
  companding 115
  dynamic range 115
  FM modulated 113
  in syncs 114
  information 21
  PPDU 141
  reserve switching flag 116
  stereo 113
Spacing attribute 63
Specification – White Book 58
Spectral colour 2
Sub-carrier
  burst 37, 41
  choice of frequency 38
  choice of frequency in PAL 49, 50
  colour 33, 41
  frequency 44
  NTSC 99
  NTSC frequencies 40
  NTSC system phase errors 42
  oscillator 37
  PAL 100
  phase locked loop 43
  SECAM 53
  SECAM deviations 53
  SECAM phase 54
Sub-code 61
Sub-page – Videotex 81
Superhet principle 89
Surface acoustic wave (SAW) filter 92
Swinging burst 49, 50
Switched mode power supply 111
Sync generator in teletext 101
Sync separator 94
Synchronisation
  of data packets 125
  of the MAC signal 121
  pulses 15, 24
  signal processing 93
  teletext bit 67
  teletext byte 67

Teledon 139, 141
Telephone line barrier 106
Telesoftware 131
  PPDU 141
Teletel 139
Teletext 58
  access time 137
  bandwidth 76
  bit rate 76
  characters 62
  code table 63
  colours 62
  conditional access 131, 134
  control codes 62
  data rate 77
  data slicer 103
  decoder 101
  display 61
  display data 71
  display memory 103
  display unit 103
  full channel 78, 136
  logic levels 74
  magazine 73
  packet 66
  page stores (memories) 78
  service 60
  World System Specification 127

Television
    colour tube   28
    control bus   110
    control unit   107
    UHF channel   22
Time compression – in the MAC
        signal   122
Transmitter – remote control   109
Tube
    camera   5
    colour television   28
Tuner unit   89
Tuning – frequency synthesis   90

U signal   33, 100
UHF television channel   22
Unused line scans   59, 60

V signal   33, 100
Varicap diode   89
Vertical insertion signals   60
Vertical resolution   19
Vestige   22
Vestigial sideband modulation   22, 25, 34
Video signal   3
    composite   15
    processing   97

Videotex   79
    CEPT   139
    charging structure   81
    code table   84
    control codes   82
    controller   106
    decoder   101
    display   82
    display memory   107
    escape sequences   83
    links   81
    page   81
    signal   83
    signal decoding   85
    signal modulation   85
    sub-page   81
    tree structure   81
Viewdata   79
Visibility function   1
Visual acuity   3, 57

Waveform – teletext raised cosine   76
Weighted
    chrominance signal in SECAM   52, 56
    colour difference signal   33, 40
White Book specification   58
White level – peak   17